EXPLO
WITH

Edward Green
Virginia Polytechnic Institute and State University

Benny Evans
Oklahoma State University

Jerry Johnson
University of Nevada, Reno

to accompany

CALCULUS

Deborah Hughes-Hallett
Harvard University

Andrew M. Gleason
Harvard University

et al.

John Wiley & Sons, Inc.
New York • Chichester • Brisbane • Toronto • Singapore

Copyright © 1994 by John Wiley & Sons, Inc.

All rights reserved.

Reproduction or translation of any part of this work beyond that permitted by Sections 107 and 108 of the 1976 United States Copyright Act without the permission of the copyright owner is unlawful. Requests for permission or further information should be addressed to the Permissions Department, John Wiley & Sons, Inc.

ISBN 0-471-09718-7

Printed in the United States of America

10 9 8 7 6 5 4 3 2 1

Table of Contents

Preface		 v
Chapter 1	A Library of Functions	 1
	•Functions •Graphs •Power functions •Exponential functions •Logarithms •Inverses of functions •Trigonometric functions •Rational functions	
Chapter 2	Key Concept: The Derivative	 61
	•Velocity •Tangent lines •Derivatives •Limits	
Chapter 3	Key Concept: The Definite Integral	 93
	•Riemann sums •Definite integrals •The Fundamental Theorem •Area •Average value	
Chapter 4	Short Cuts to Differentiation	 119
	•Difference quotients •Differentiation rules •Implicit differentiation	
Chapter 5	Using the Derivative	 133
	•Maxima and minima •Inflection points •Applications •Newton's method	
Chapter 6	Reconstructing a Function from its Derivative	 163
	•Antiderivatives •Graphing antiderivatives	
Chapter 7	The Integral	 169
	•Approximating integrals •Improper integrals	
Chapter 8	Using the Definite Integral	 205
	•Applications •Arc length •Volume •Distributions	

Chapter 9   Differential Equations 229

       •Families of solutions •Slope fields •Euler's method •Applications of first-order equations •Equilibrium solutions •Systems of equations •Second-order equations

Chapter 10   Approximations 265

       •Taylor polynomials •Interval of convergence •Fourier series

Appendix I   *Mathematica* ® Reference and Tutorials 287

Appendix II   Code for a *MATHEMATICA* Package 307

Index of Solved Problems 315

Index of Laboratory Exercises 319

PREFACE

This manual is an enrichment supplement to the *Calculus* text by Hughes-Hallett, Gleason et al., which was the result of a major NSF initiative in calculus reform called **The Calculus Consortium based at Harvard**. We will refer to it throughout the book as the **CCH Text** for short. This manual was first developed in conjunction with the computer program Derive ®. The present book basically translates that manual for use with *MATHEMATICA*. The Derive manual was written by Benny Evans and Jerry Johnson and the revision to *MATHEMATICA* was done by Ed Green.

The purpose of this manual is to help students use *Mathematica* ® as a tool to explore Calculus in the spirit of the CCH Text, beyond the level of rote calculations and superficial exercises. Most of the problems are different from those which one would normally expect to do with nothing but pencil and paper.

Some authors have taught from the CCH Text using *MATHEMATICA* and have found that they make a marvelous pair. Although this manual is written as a supplement to the CCH Text, it may be used with any calculus text as a source of stimulating problems.

We would like to think that many of the exercises in the earlier manual *Discovering Calculus with Derive* ®, also published by John Wiley & Sons, embody much the same spirit as the CCH Text, but this manual is much more focused, far leaner, and is written specifically for CCH Text.

In the spirit of the CCH Text, this manual is written for students – not instructors. One common thread running through it and the CCH Text is the persistent admonition, "EXPLAIN YOUR ANSWER." Written explanations in clear, complete English sentences are assumed to be part of all assignments.

Mathematica is a registered trademark of Wolfram Research, Inc.
To order *MATHEMATICA* contact Wolfram Research, Inc., Champagne, Illinois 61824
Telephone: 800–441–6284   Fax 217–398–0747

Key Features

- No prior knowledge of *MATHEMATICA* is required. Appendix I summarizes important commands and will help novice users get started, but students with a little *MATHEMATICA* experience should be able to begin in Chapter 1 with no difficulty.

- The book is written with the reform calculus spirit expressly in mind. Accordingly, the problems go beyond the level of rote calculations and "template" exercises.

- Optional code for creating graphics and automating some calculations is provided in Appendix II.

- Exercises are designed as laboratory sheets that students may detach and hand in.

- Most of the problems have been class tested by the authors and others.

MATHEMATICA is very impressive, but our central purpose is to teach *mathematics*, not to show off hardware and software. We have tried to include enough *MATHEMATICA* instructions and suggestions in both the Solved Problems and Appendix I so the reader will not have to spend a great deal of time referring to the *MATHEMATICA* manual, but this is not intended as a substitute for it.

Structure of the Book

The chapters are designed to follow the CCH Text. Each chapter starts with a set of Solved Problems followed by a set of Laboratory Exercises. The Solved Problems are examples that provide a context for *MATHEMATICA* instructions that one is likely to encounter in the exercises, but they also contain useful mathematics hints and lessons as well.

The Laboratory Exercises are sets of problems that students should be able to do upon completion of the appropriate Solved Problem. They are sufficiently complex that solving them without assistance from a computer or graphing calculator is usually not practical. Most of them invite the student to discuss their observations and findings. Each Solved Problem and each Laboratory Exercise contains a reference to the section in the CCH Text where it fits.

The *MATHEMATICA* Program

> *The novice who has never used MATHEMATICA before should begin by reading Sections 1 and 2 of Appendix I.*

MATHEMATICA is a comprehensive computer algebra system that runs on PC and Macintoshes, as well as larger computers. We assume that you are using either a Macintosh or a PC with Windows. The user interfaces on these two platforms are slightly different from one another. A student using either a Macintosh or a PC will be able to use this book. Warning: The actual output you obtain with your computer may differ slightly from that shown in the text, depending on the computer you are using.

Any sophisticated software takes some practice and experience to master, but we have used *MATHEMATICA* in our courses for the last few years and are convinced that one of its strongest educational points is that it is easy for students to learn and to use. It is also powerful enough for professional applications in mathematics, science, and engineering.

This manual refers to *MATHEMATICA* version 2.2, and all figures were created using that version of the program. We have not tried to elaborate its differences from earlier versions of *MATHEMATICA*. (For example, version 2.2 is the first to work with Windows.)

Suggestions for Incorporating *MATHEMATICA*: What Has Worked.

- We normally spend a period with our classes at the computer laboratory in the first week or two of the semester to introduce them to *MATHEMATICA*. Our experience is that students quickly become comfortable with *MATHEMATICA*, and that only minimal help is needed later.

- Since the students find the syntax of *MATHEMATICA* difficult, we have found that introducing new commands as they are needed is most successful. We spend a few minutes discussing the syntax and giving examples each time a new command is introduced.

- After this initial session we may return to the lab with the class two or three more times during the semester. It is important to try to integrate the lab experience with material that is currently being covered in the course. We often begin such a lab session by asking students to work through a Solved Problem and later give them a quiz that consists of a similar exercise. Grades in the lab are important. Even the best students appreciate credit for their work, and there are always students who will remain passive in the lab without the incentive of a grade.

- We give outside lab assignments from time to time. The frequency may vary from five to ten assignments a semester. We encourage students to work together on the assignments, but we insist that they turn in their own printouts and write up the results in their own words.

- We have found that some classroom discussion of an assigned exercise is desirable, including a few words about new *MATHEMATICA* commands. However, the **Solved Problems** and Appendix I should contain the essentials.

- To avoid extra expense to the student, a school's lab might buy enough manuals for one class (say 30 or 40) and check them out to students during lab time.

Important Conventions

1. *MATHEMATICA* is either menu driven or driven by keyboard commands. Since the menus on the Macintosh and PC notebooks differ slightly, we have given the keyboard commands when possible. As students become familiar with the particular machine on which they are running *MATHEMATICA*, they learn how to use the menu commands not described in this book. Most of these commands are self-explanatory.

2. When you must enter an expression into *MATHEMATICA* it will be displayed in **boldface**. In this case, you type exactly the symbols you see except the period at the end. For example, if we want you to enter $\frac{x^2}{3}$, we will say evaluate **x^2/3**. To "evaluate" a command on a PC, press the Shift key and the Enter key simultaneously. On a Macintosh, press the Ctrl key and the Return key simultaneously, or press the Enter key. On either platform, you can evaluate an entry using the mouse and the menu activated from the Action icon. This is described in Chapter 1.

A Word to Students

MATHEMATICA has essentially automated most of the standard algebra calculations you normally encounter, just as scientific calculators have done with arithmetic. It will simplify complicated expressions, solve equations, and draw graphs. It will also do most of the standard calculations that arise in calculus such as finding derivatives and integrals. But that doesn't mean calculus and algebra are obsolete or unimportant. Even though calculators will do arithmetic, we still have to know what questions to ask, understand what the answers mean, and realize when an obvious error has been made. In the same way, we still have to understand the definitions, concepts, and processes that are involved in algebra and calculus so we will know what to tell *MATHEMATICA* to do, understand the answers, and be able to detect errors. *MATHEMATICA* only does the calculations; you must still do the thinking.

Always view any computer or calculator output critically. Be alert for answers that seem strange – you might have hit the wrong key, entered the wrong data, or made some other mistake. It is even possible that the program has a bug! If a problem asks for the cost of materials to make a shoe box and you get \$123.28 or −\$0.55 you should suspect something is wrong!

Clear communication is *at least* as important in mathematics as in other fields. You should always write your answers neatly in complete, logical sentences. Reread what you have written and ask yourself if it really makes sense.

Before you turn on the computer, work through as much of an assigned problem as you can with pencil and paper, taking note of exactly where you think the computer will be required and for what purpose. You may be surprised at how little time you will actually have to spend in front of the machine if you follow this advice.

Final note

The code for a *MATHEMATICA* package called MYCALC.m is given in Appendix II. Instead of having to type this code, it is available via anonymous ftp. To obtain the details on how to get the code, contact Ed Green, Department of Mathematics, Virginia Tech, Blacksburg, VA. 24061. Internet email address: green@math.vt.edu

---

Ed Green

July 1994

Chapter 1
A Library of Functions

Two of the features of *MATHEMATICA*, plotting and solving equations, will be used extensively in this course. The necessary *MATHEMATICA* instructions are included in the **Solved Problems** that follow, but you may find it helpful to refer to Sections 3 and 4 in Appendix I for an overview of plotting and Sections 7 and 8 of Appendix I for equation solving.

Solved Problem 1.1: Domains, ranges, and zeros of functions (CCH Text 1.1)

Plot the graph and find the domain, range, and zeros of each of the following functions.

Remark: A formal discussion of "zeros" in the CCH Text is deferred to Appendix A. However, we will include the topic here because it arises throughout the text (see 5 (a) in the Review Problems for Chapter 1).

(a) $f(x) = \dfrac{x^3 - 7x^2 - x + 7}{50}$

(b) $f(x) = x^4 - \sqrt{1-x}$

(c) $f(x) = \sqrt{1-x} + \sqrt{x-1} + 0.5$

Solution to (a): First, we define the function $f(x)$. Type **Clear[f]**. This clears any previous definition of f. Evaluate this either by opening the Action menu and selecting the Evaluate Selection option or by pressing the Shift key and the Enter key simultaneously. Henceforth, we will just say "evaluate the input" or simply "evaluate," for either of these options. Because you just cleared any previous definition of f, you will see no output. Now type **f[x_] := (x^3 - 7x^2 -x +7)/50** and evaluate this input. Again, you will not see any output; but you have now defined the function $f(x)$. To graph the function, type **Plot[f[x],{x,-5,5}]** and evaluate this input. This plots the graph of $y = f(x)$ from $x = -5$ to $x = 5$. *MATHEMATICA* produces output that should look similar to Out[3] in Figure 1.1. In algebra we learn that a cubic polynomial may have as many as three roots. Are there really only two, as shown in Out[3] of Figure 1.1? We chose to plot $f(x)$ from $x = -5$ to $x = 5$. Let's make the domain interval larger. To do this, either type **Plot[f[x],{x,-10,10}]** or edit the input line **Plot[f[x],{x,-5,5}]** by changing -5 to -10 and changing 5 to 10. Then evaluate this input. You now get the graph shown in Out[4] of Figure 1.1. The graph can be seen crossing the x axis at three points. Thus, there are indeed three zeros. Furthermore, this picture suggests that the domain and range of f are both $(-\infty, \infty)$.

1

```
In[1]:=
   Clear[f]
In[2]:=
   f[x_]:=(x^3-7x^2-x+7)/50
In[3]:=
   Plot[f[x],{x,-5,5}]
```

```
In[4]:=
   Plot[f[x],{x,-10,10}]
```

```
Out[4]=
   -Graphics-
In[5]:=
   Solve[f[x]==0,x]
Out[5]=
   {{x -> -1}, {x -> 1}, {x -> 7}}
```

```
Out[3]=
   -Graphics-
```

Figure 1.1: **The graph of** $\dfrac{x^3 - 7x^2 - x + 7}{50}$

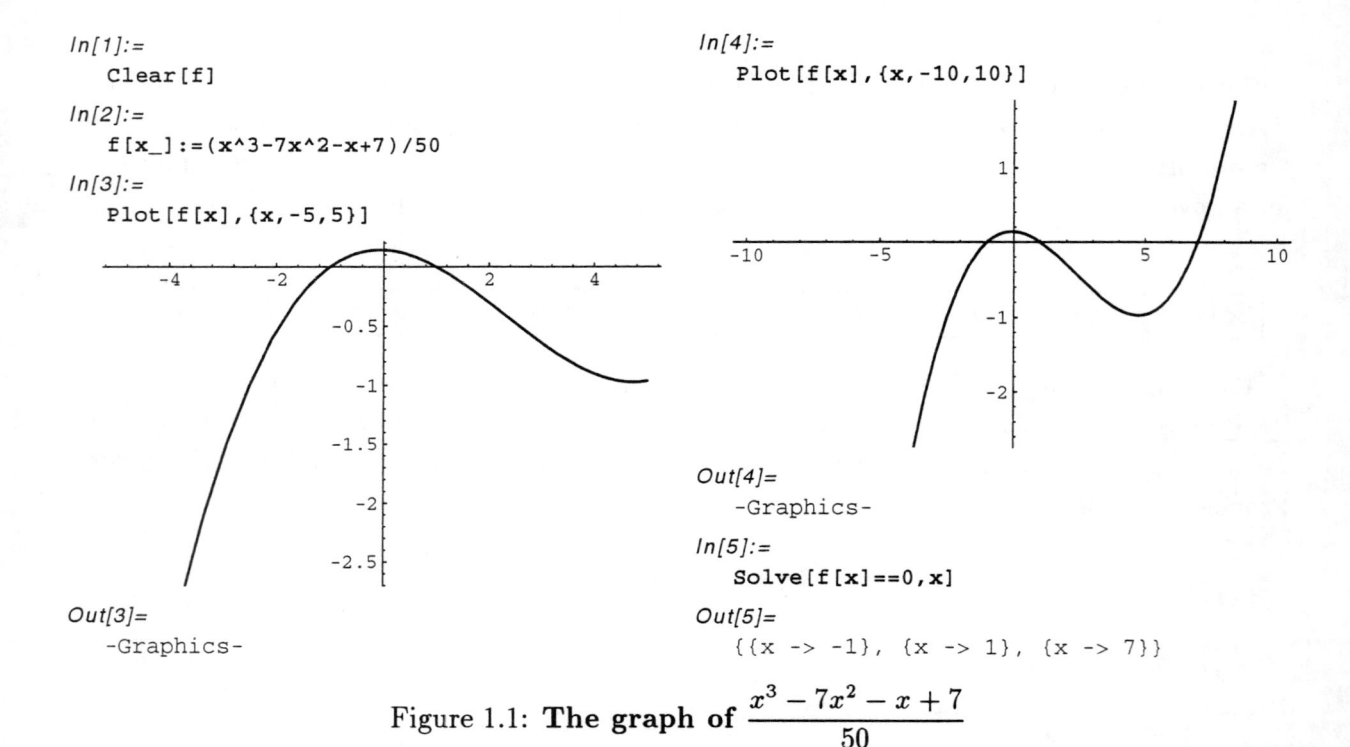

Remark: Graphing programs may not initially display all the interesting information about a function, e.g., all the zeros, and some extra exploration may be required. The moral to the story is to always think critically about everything in general, and computer output in particular.

Now we may use the *MATHEMATICA* function Solve to find the zeros. Type **Solve[f[x]==0,x]** and evaluate. We see the three zeros in Out[5] of Figure 1.1.

Solution to (b): Evaluate the following three inputs: **Clear[g]**, **g[x_]:= x^4-Sqrt[(1-x)]**, and **Plot[g[x],{x,-1,1}]** as in Part (a), and the graph appears as in Out[3] of Figure 1.2. It is easy to see what the domain of $x^4 - \sqrt{1-x}$ is without a computer! $1 - x$ must be nonnegative to avoid taking the square root of a negative number. Hence, the domain consists of all real numbers x that are less than or equal to 1, and the graph confirms this. If you try to plot $g(x)$ for values $x > 1$, *MATHEMATICA* will give you error messages. It is not so easy to see what the range is. The graph in Figure 1.2 indicates that function values may be arbitrarily large but can be no smaller than the "bottom" of the graph. We need the y coordinate for this spot. We use a

MATHEMATICA graphics feature for this. Move the mouse pointer into the graphics window. Clicking the mouse button once causes a border to appear. In the graphics window, move the mouse pointer to one of the darkened rectangles on the border so that the pointer changes to <->. While depressing the mouse button, move the mouse to resize the graphics window. See Out[4]. To find the coordinates of a point in a graphics window, move the mouse pointer into the window and click the button once. Hold down the Ctrl key on a PC, or the Command and Apple keys on a Macintosh, click the mouse button and the coordinates of the mouse pointer can be found on the lower left-hand side of the bottom border of the *MATHEMATICA* window. Depressing the Ctrl key or the Command-Apple keys and moving the mouse gives other coordinates. Doing this for our graph, we see the coordinates are approximately $(-0.486, -1.16)$. If we used a small domain interval around -0.486, say $[-.49, -0.45]$, and replotted g we could get a better approximation. Thus the range of f is approximately $[-1.16, \infty)$.

Let's find the zeros. If we ask *MATHEMATICA* to **Solve** as we did in Part (a), it will return

Out[7]= {ToRules[Roots[x + x^8 == 1, x]]}. This indicates that *MATHEMATICA* cannot find the zeros of this function using Solve, but we can still get approximate values using the *MATHEMATICA* function called FindRoot. Use the graph to find x values near the zeros of the function. We can use $x = -1$ and $x = .7$. Type **FindRoot[g[x]== 0,{x,-1}]**. The output gives an approximation of the x value. Similarly, **FindRoot[g[x] == 0,{x,.7}]** approximates the other zero.

You may be interested to know that the *MATHEMATICA* FindRoot function uses a technique known as the *Newton's method* to approximate zeros. This is discussed in Section 5.7 of the CCH Text.

Solution to (c): This is tricky, but it has a serious point. We evaluate **Clear[h]**, then **h[x_] := Sqrt[1-x]+Sqrt[x-1]+0.5**. When we **Plot** it as we did with (a) and (b) above, we get error messages and no graph. What happened?

For $\sqrt{1-x}$ to be defined, we must have $x \leq 1$. For $\sqrt{x-1}$ to be defined, we must have $x \geq 1$. Therefore, the domain of $\sqrt{1-x} + \sqrt{x-1} + 0.5$ consists of a single number, 1, and consequently the range consists of a single number, 0.5. There are, of course, no zeros of this function. <u>Remark</u>: We didn't need a computer for this. As we often stress, one should think about a problem before trying to use a computer to solve it.

The graph of this function consists of the single point $(1, 0.5)$, which cannot be easily plotted by *MATHEMATICA*.

The steps used in this problem to make graphs and to solve equations are so important that they are worth summarizing again.

In[1]:=
    Clear[g]
In[2]:=
    g[x_]:= x^4-Sqrt[1-x]
In[3]:=
    Plot[g[x],{x,-1,1}]

Out[3]=
    -Graphics-

In[4]:=
    Plot[g[x],{x,-1,1}]

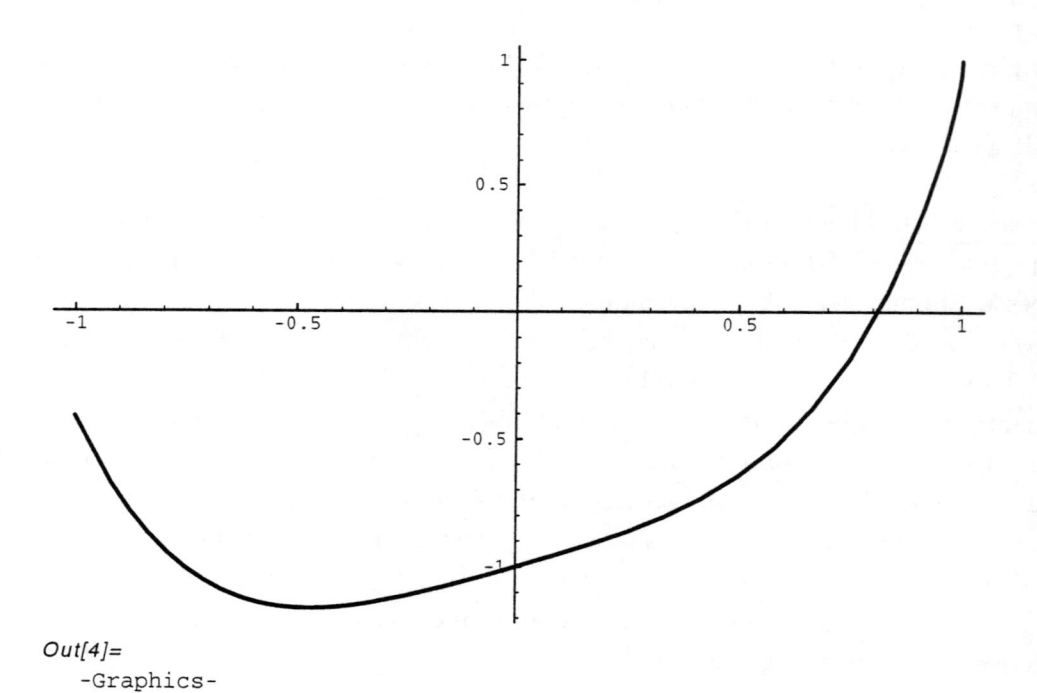

Out[4]=
    -Graphics-

Figure 1.2: **Graph of** $x^4 - (1-x)^{\frac{1}{2}}$

PLOTTING GRAPHS

Step 1: Use the Clear[ ] function to remove any previous definitions.

Step 2: Enter the function definition, which has the form $f[x\_] := \cdots$ (Note that you can use any variable you want in place of x.)

Step 3: Choose an interval, say $[a, b]$, for the plot.

Step 4: Type **Plot[f[x],{x,a,b}]**, and evaluate to produce the graph.

For further details on graphs consult Sections 3 and 4 of Appendix I.

APPROXIMATE SOLUTIONS OF EQUATIONS

For equations that *MATHEMATICA* cannot solve using the **Solve** function, proceed as follows.

Step 1: Plot the function to find a value a such that $f(a)$ is near 0.

Step 2: Type **FindRoot[f[x],{x,a}]** and evaluate to get the approximate root near $x = a$. Warning: If you get a message that Newton's method did not converge after 15 steps, try to find a better value for a. This can be done by replotting the graph after changing the domain values closer to the point of interest.

Step 3: Use the graph to find other values a close to other roots of the equation, if such roots exist.

Refer to Sections 7 and 8 of Appendix I for further information on solving equations.

Laboratory Exercise 1.1

Zeros, Domains, and Ranges (CCH Text 1.1)

Name _____   Due Date _____

Plot the following functions. Find the zeros, domain, and range of each. Explain how you obtained your answers.

1. $f(x) = 3 + 15x^2 - x^4$

2. $g(x) = \sqrt{2x-1} + \sqrt{x-8} - x$

3. $h(x) = x^5 - x^3 - 900x + 901$. (You may need to try different domain intervals to get a good picture.)

4. $i(x) = 0.25 + \sqrt{6-x} - \sqrt{x-6}$

5. $j(x) = \sqrt{-2x^4 + 9x^3 - 6x^2 - 11x + 6}$. (<u>Hint</u>: Factor $-2x^4 + 9x^3 - 6x^2 - 11x + 6$ using *MATHEMATICA's* Factor command.)

Solved Problem 1.2: Testing exponential data (CCH Text 1.3)

An experiment yields the following data:

$$(1,\ 0.075),\ (3,\ 0.36),\ (5,\ 1.75),\ (7,\ 8.48),\ (9,\ 41.05)$$

(a) Show that the data can be approximately fit by an exponential function, $y = ab^x$.

(b) Find the specific formula that fits the first two data points.

(c) Plot the data points and the exponential function on the same screen to confirm your findings.

Solution to (a): The basic principle is that an exponential function is one for which ratios of successive function values, with equal changes in the variable, are the same. Let's test the data by computing these ratios:

$$\frac{0.36}{0.075} = 4.8,\quad \frac{1.75}{0.36} = 4.861,\quad \frac{8.48}{1.75} = 4.846,\quad \frac{41.05}{8.48} = 4.841$$

These results suggest that the data are approximately exponential.

Solution to (b): Now that we know the data fit the formula $y = ab^x$, we must find the numbers a and b. We can assert that $0.075 = ab^1$ and $0.36 = ab^3$. Dividing the second equation by the first gives $\frac{0.36}{0.075} = \frac{ab^3}{ab}$. This simplifies to $4.8 = b^2$, so $b \approx 2.19$.

We use the fact that $0.075 = ab$ to obtain $a = \frac{0.075}{b} \approx \frac{0.075}{2.19} \approx 0.0342$, and conclude that the data approximately fit the function $y = (0.0342)\, 2.19^x$.

Solution to (c): Now we must make a plot. Here is where *MATHEMATICA* will be a big help. To plot data points, we use *MATHEMATICA*'s ListPlot command. The ListPlot command is discussed in Section 3 of Appendix I. First, remove any previous definition of T with the command **Clear[T]**. Then create a list of points by typing:

```
T = {{1, 0.075}, {3, 0.36}, {5, 1.75}, {7, 8.48}, {9, 41.05}}
```

Notice that we used *set brackets* surrounding the pairs and the *set brackets* for the ordered pairs. The result is seen as a *list* in Out[2] Figure 1.3.

Now, evaluate **ListPlot[T]** to get a plot of these points. To force a specific origin for the axes, use the AxesOrigin command. Evaluate **ListPlot[T, AxesOrigin -> {0,0}]** to have the axes cross at the origin. Use the PlotRange command to obtain a specific range in graphics. Evaluate **ListPlot[T, PlotRange -> {0,10}]**

```
In[1]:=
   Clear[T]
In[2]:=
   T={{1,0.075},{3,0.36},{5,1.75},{7,8.48},
   {9,41.05}}
Out[2]=
   {{1, 0.075}, {3, 0.36}, {5, 1.75}, {7, 8.48},
   {9, 41.05}}
In[4]:=
   ListPlot[T]
Out[4]=
   -Graphics-
In[5]:=
   Clear[f]
In[6]:=
   f[x_]:=(0.0342)*(2.19)^x
In[7]:=
   Show[ListPlot[T],Plot[f[x],{x,0,10}]]
Out[7]=
   -Graphics-
```

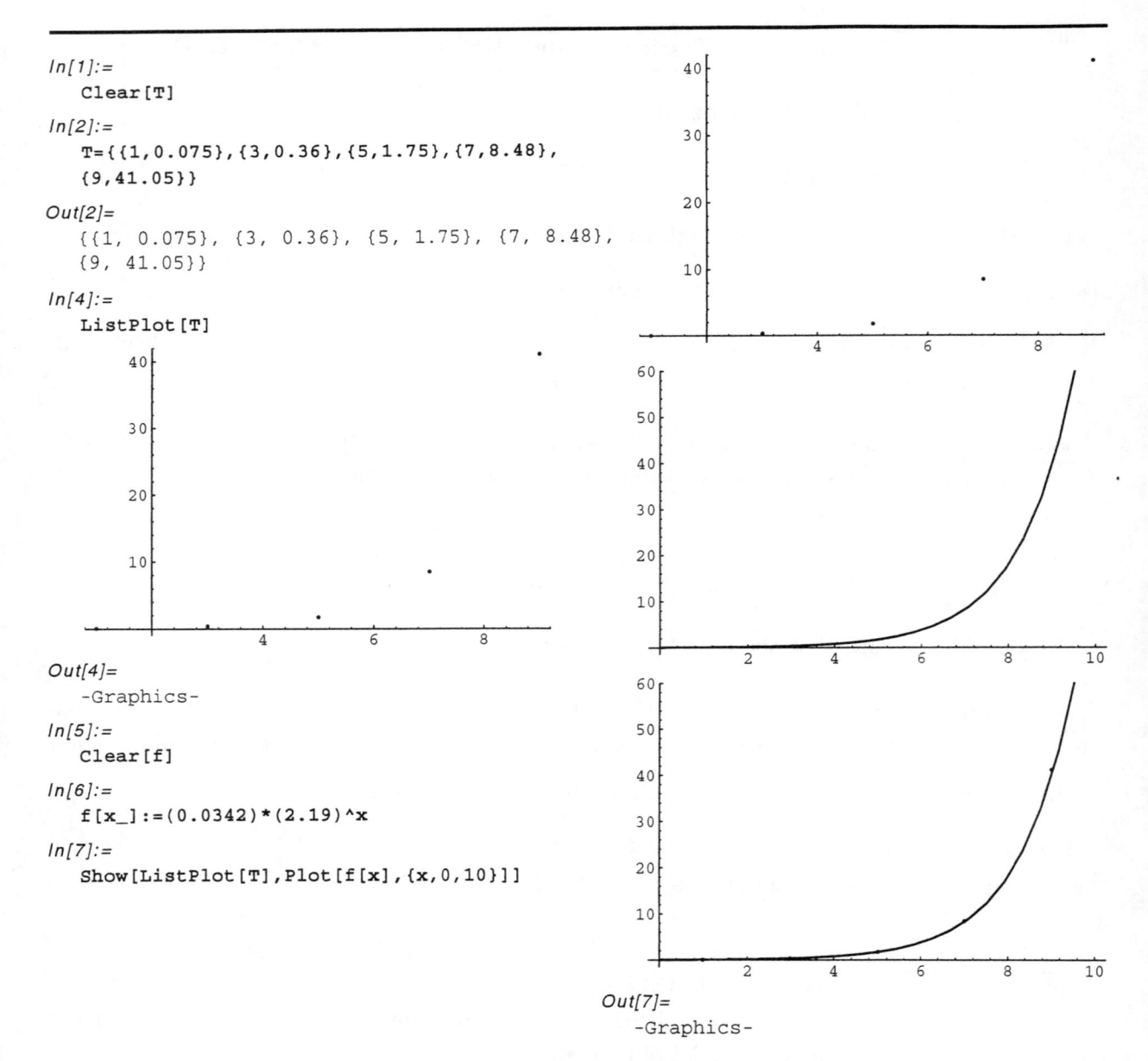

Figure 1.3: **Plot of data and** $(0.0342)2.19^x$

To show both the exponential function and points, we define the exponential function. Evaluate **Clear[f]** and $f[x\_] = (0.0342)*(2.19)\hat{}x$. Now we use the Show command to see both the points and the function together. Evaluate **Show[ListPlot[T],Plot[f[x],{x,0,10}]]** to get the graphs above Out[7] in Figure 1.3. We can see that the curve goes nicely through all the data points, confirming our calculations.

Laboratory Exercise 1.2

Fitting Exponential Data (CCH Text 1.3)

Name _____   Due Date _____

An experiment yields the following data:

$$(-2.5,\ 5.6),\quad (-0.5,\ 2.75),\quad (1.5,\ 1.35),\quad (3.5,\ 0.66),\quad (5.5,\ 0.32)$$

1. Show that the data can be approximately fit by an exponential function, $y = ab^x$.

2. Find the specific formula that approximately fits the data.

3. Plot the data points and the exponential function to confirm your findings. Be sure your plot shows all five data points.

Laboratory Exercise 1.3

U. S. Census Data (CCH Text 1.3)

Name _____   Due Date _____

This exercise refers to the table of census data in problem 16 of Section 2.4 in the CCH Text. To make a more readable plot, you should modify these data by taking the year 1790 as zero, 1800 as 10, etc. Thus, the first two points will be (0, 3.9) and (10, 5.3), and the last one will be (200, 226.5).

1. Find the specific exponential formula, $y = ab^x$, that fits the first two data points.

    Plot the exponential function and all 21 of the modified data points together. Be sure your plot shows all the data points.

2. Estimate the first decade during which the curve and the data diverge. What historical event may have influenced this divergence?

3. If the population continued to grow at the same rate as it did during the period from 1790 to 1800, what would it have been in 1980? What would it be today?

4. Find the specific exponential formula, $y = ab^x$, that fits the decade 1890 to 1900. Plot this exponential function and all 21 of the modified data points together.

5. What are the two decades during the twentieth century where the data and the curve in Part 4 first diverge? What historical events may have influenced this divergence?

6. If the population continued to grow at the same rate as it did during the decade 1890 to 1900, what would it have been in 1980? What would it be today?

Solved Problem 1.3: Powers versus exponentials (CCH Text 1.4)

Plot the graphs of x^6 and 6^x. Determine the points where they intersect. For what values of x is $6^x > x^6$?

<u>Solution</u>: We evaluate **Clear[f,g]** and then create the two functions by evaluating **f[x_] := x^6** and **g[x_] := 6^x**. See Out[1]-[3] in Figure 1.4. Then evaluate **Plot[{f[x],g[x]},{x,-1,1}]** as in Out[4] of Figure 1.4. It is hard to tell which curve is which. To help with this, you may use the Thickness option in PlotStyle. Evaluate **Plot[{f[x],g[x]},{x,-1,1},PlotStyle -> {Thickness[.001],Thickness[.01]}]** to obtain Out[5] in Figure 1.4. The thicker curve is $g(x)$. If you have a color monitor, you can use the RGBColor command instead of the Thickness command. The RGBColor command is described in Section 3 of Appendix I.

We can see that the graphs cross between $x = -1$ and $x = 0$. Following the technique presented in Part (b) of Solved Problem 1.1 we can find the crossing point accurately. Evaluate **FindRoot[f[x]==g[x],{x,-.5}]**. The result, $x = -0.789877$, appears in Out[6] of Figure 1.4.

We can see from Out[5] that the two graphs seem to be coming back together on the positive $x$axis. If we evaluate **Plot[{f[x],g[x]},{x,1,2}]** we see a second crossing point in Out[7] of Figure 1.4 at approximately $x = 1.6$. We evaluate **FindRoot[f[x]==g[x],{x,1.6}]** to obtain the more accurate value, $x = 1.62424$, in Out[8] of Figure 1.4.

We do not need a computer to tell us that the graphs cross once more at $x = 6$. Thus $x^6 < 6^x$ on the interval $(-0.789877, 1.62424)$ and again on $(6, \infty)$.

Both of these graphs grow so rapidly that it is difficult to display all their interesting features on a single screen. In Out[1] of Figure 1.5 we have set the domain to be $(-1, 10)$ to show the crossing at $x = 6$. Notice that with this scale, the other two crossings cannot be seen.

This illustrates the fundamental concept that for bases greater than 1, exponential functions grow faster than power functions. That is, *if x is sufficiently large, then an exponential (a^x with $a > 1$) will eventually be larger than a power (x^b)*. So 6^x is destined to eventually remain on top of x^6.

In[1]:=
```
Clear[f,g]
```

In[2]:=
```
f[x_]:=x^6
```

In[3]:=
```
g[x_]:=6^x
```

In[4]:=
```
Plot[{f[x],g[x]},{x,-1,1}]
```

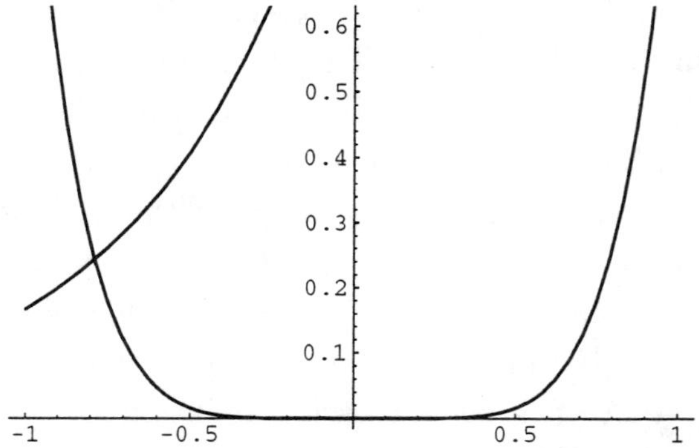

Out[4]=
  -Graphics-

In[5]:=
```
Plot[{f[x],g[x]},{x,-1,1},PlotStyle->
  {Thickness[0.001],Thickness[0.01]}]
```

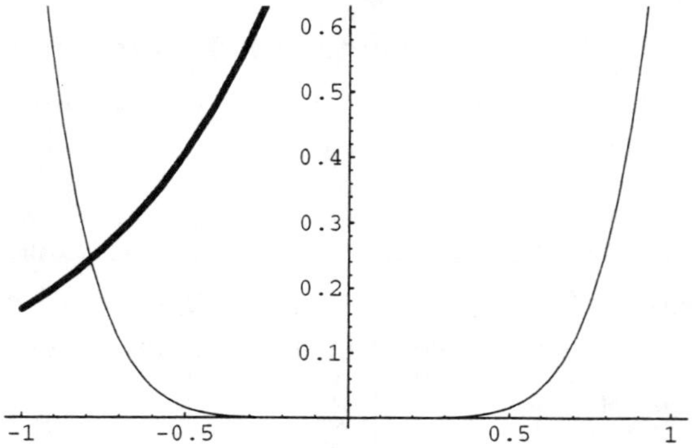

Out[5]=
  -Graphics-

In[6]:=
```
FindRoot[f[x]==g[x],{x,-.5}]
```

Out[6]=
  {x -> -0.789877}

In[7]:=
```
Plot[{f[x],g[x]},{x,1,2}]
```

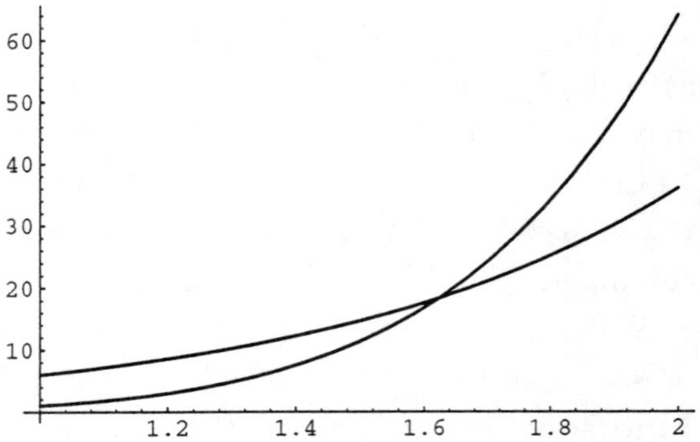

Out[7]=
  -Graphics-

In[8]:=
```
FindRoot[f[x]==g[x],{x,1.6}]
```

Out[8]=
  {x -> 1.62424}

Figure 1.4: **Plot of 6^x and x^6**

16

In[1]:=
    Plot[{x^6,6^x},{x,-1,10}]

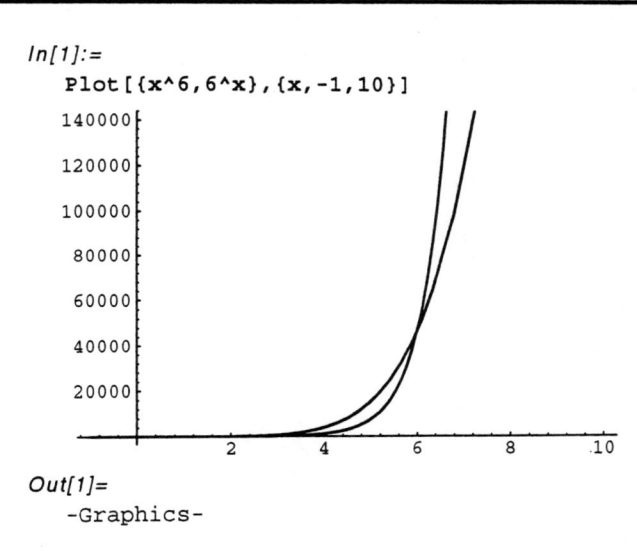

Out[1]=
    -Graphics-

Figure 1.5: 6^x **and** x^6: **domain** $(-1, 10)$

Laboratory Exercise 1.4

Powers Versus Exponentials (CCH Text 1.4)

Name _____   Due Date _____

1. Plot the graphs of x^4 and 3^x. Determine the points where they intersect correct to six decimal places.

2. For what values of x is $3^x > x^4$?

3. Plot the graphs of x^5 and 2^x. Determine the points where they intersect correct to six decimal places.

4. Take the largest intersection point you found in Part 3 above and substitute it for x into $2^x - x^5$. What do you get? $2^x - x^5$ *should* be zero there. Is it close to zero? Explain what's going on.

5. For what values of x is $2^x > x^5$?

Solved Problem 1.4: Inverses of functions (CCH Text 1.5)

Decide whether the following functions have inverses. For those that have inverses, (i) calculate $f^{-1}(4)$ and $f^{-1}(2)$, and (ii) plot both f and its inverse on the same coordinate axes.

(a) $f(x) = \dfrac{x^4}{20} - 5x$

(b) $f(x) = x^5 + 2x + 1$

<u>Solution to (a)</u>: If we evaluate the following three inputs, **Clear[f]**, **f[x_] := (x^4)/20 - 5x**, and **Plot[f[x],{x,-1,1}]**, we get the graph above Out[3] of Figure 1.6. It looks as if it's decreasing and passes the horizontal line test, so we might be tempted to say it's invertible. But if we increase the domain to $(-5, 5)$, we see the result in Out[4] of Figure 1.6. This graph fails the horizontal line test (notice that the x axis crosses the graph twice) and thus is not invertible. The original choice of domain $(-1, 1)$ was too small to reveal enough of the graph.

In[1]:=
   Clear[f]
In[2]:=
   f[x_] := (x^4)/20 - 5x
In[3]:=
   Plot[f[x],{x,-1,1}]

In[4]:=
   Plot[f[x],{x,-5,5}]

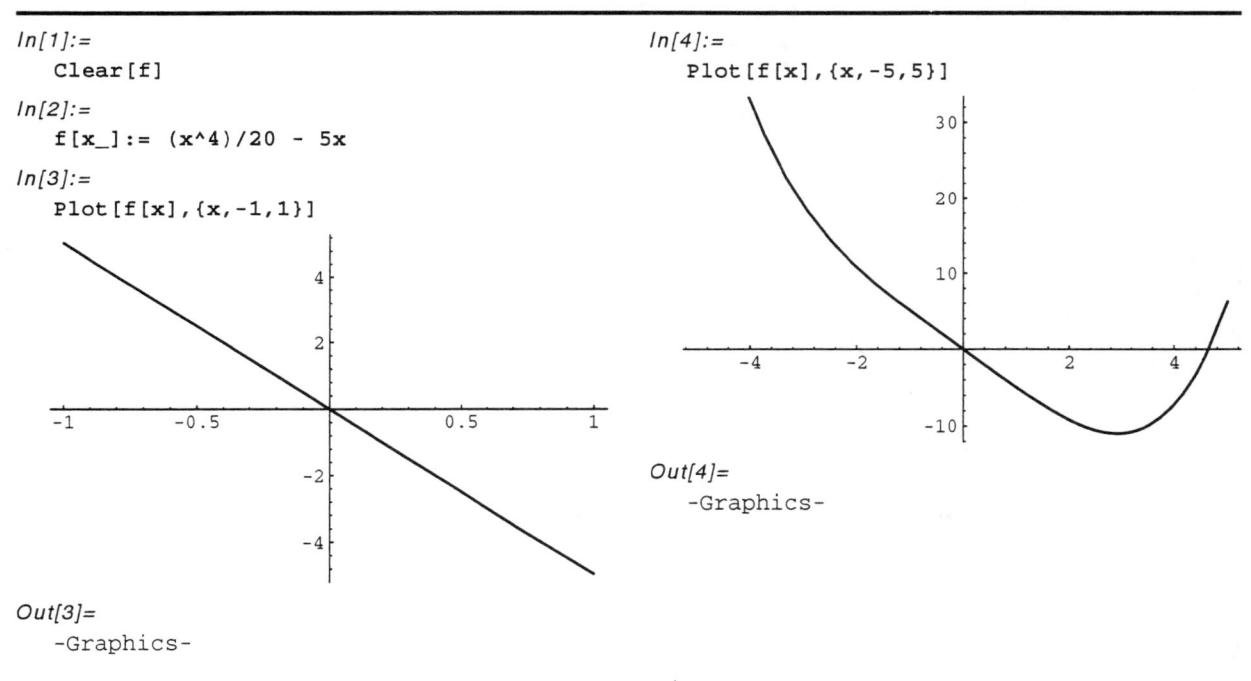

Out[4]=
   -Graphics-

Out[3]=
   -Graphics-

Figure 1.6: **The graph of** $\dfrac{x^4}{20} - 5x$ **with domains**

<u>Remark</u>: As we often stress, one must not accept computer output uncritically. In Section 1.11 of the CCH Text, we will see that $\dfrac{x^4}{20} - 5x$ goes to ∞ as $x \to \infty$. Therefore it must eventually turn around and increase, which means it will fail the horizontal line test. We reason this without using a computer graph.

Solution to (b): We evaluate **Clear[f]** and **f[x_] := x^5+2x+1** as seen above Out[31] of Figure 1.7. Increasing functions have inverses. So do decreasing ones. That's because the equation $f(x) = a$ cannot have more that one solution for x in these cases. (Explain why.) Our function is increasing (see Section 1.3 of the CCH Text to recall the definition), as we can see by plotting. (Which of the two graphs in Figure 1.7 is it?)

In[20]:=
    `Clear[f]`
In[21]:=
    `f[x_]:=x^5+2x+1`

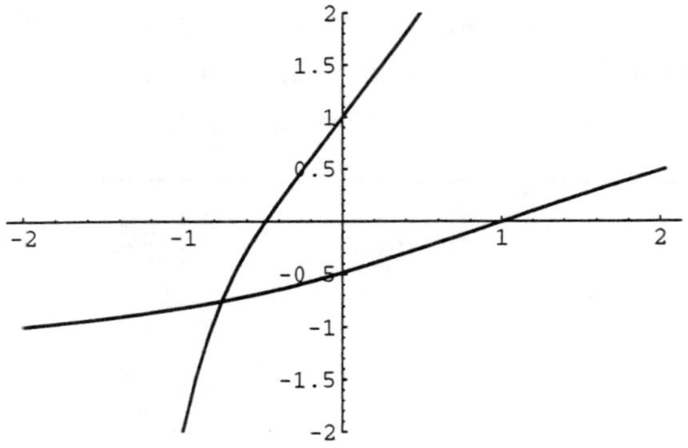

Out[31]=
    -Graphics-
In[32]:=
    `FindRoot[f[x]==2,{x,1.5}]`
Out[32]=
    `{x -> 0.486389}`

Figure 1.7: **The graphs of f and f^{-1}**

As we just saw in Part (a), we must be careful not to fall into a trap. The evidence provided by the graph in Figure 1.7 does not guarantee that it won't "wiggle" somewhere off the screen. We can verify that f is increasing by showing with algebra that if $x_1 < x_2$, then $f(x_1) < f(x_2)$. We leave it to the reader as an exercise to do this. (Argue first that if $x_1 < x_2$, then $x_1^5 < x_2^5$ and $2x_1 < 2x_2$. We reiterate: This argument does not use a computer.)

Solution to (b) Part (i): It is not possible to find a general formula for the inverse of f, but we can calculate the special cases, $f^{-1}(4)$ and $f^{-1}(2)$.

If $x = f^{-1}(4)$, then $f(x) = 4$. Thus, we need to solve the equation $x^5 + 2x + 1 = 4$. It's easy to see that $x = 1$ is a solution without a computer to help us, so we can conclude $f^{-1}(4) = 1$.

If $x = f^{-1}(2)$, then $f(x) = 2$. Thus, we need to solve the equation $x^5 + 2x + 1 = 2$. Since $f(x) = x^5 + 2x + 1$ is already defined, we can try to evaluate **Solve[f[x] == 2,x]**. *MATHEMATICA* cannot solve this equation exactly, so we will have to find an approximate solution. From the graph of f above Out[31] of Figure 1.7 we see that $f(x) = 2$ somewhere between $x = 0$ and $x = 1$. Thus, we evaluate **FindRoot[f[x]==2,{x,1.5}]** From Out[32] of Figure 1.7 we see that $f^{-1}(2) \approx 0.486389$.

<u>Solution to (b) Part (ii)</u>: <u>Discussion</u>: Since we can't get an explicit formula for $f^{-1}(x)$, how can we graph it? The graph of a function f with domain D is the set of ordered pairs $(x, f(x))$ with x in D. Thus if we ask *MATHEMATICA* to **Plot** the function f, we get the same result as plotting the ordered pairs **[x,f(x)]**. We can plot such pairs by using the *MATHEMATICA* commands ListPlot and Table. Try evaluating **ListPlot[Table[{x,f[x]},{x,-1,1,.01}]]**. The entry .01 is the step that x is being evaluated from -1 to 1. (The Table command is discussed in Section 13 of Appendix I.) If we used .1 instead of .01, many fewer points would be plotted.

A better method is to use the ParametricPlot command. Viewing x as a parameter, we want to plot the points of the form $(f(x), x)$. If f has an inverse, then the plot of the ordered pairs $(f(x), x)$ is the graph of f^{-1}.

We now explain how we got the inverse graph. We evaluate the following three inputs: **Clear[plotf,plotinv]**, **plotf = Plot[f[x],{x,-1,1,.01}]]**, and **plotinv = ParameticPlot[{f[x],x},{x,-1,0.5}]]**. Next we use *MATHEMATICA*'s Show command, which displays graphics together. Evaluate **Show[plotf,plotinv]** to get both graphs as desired in Out[31] of Figure 1.7.

We take note of the fact that the graphs are mirror images of each other through the line $y = x$. (You might find it instructive to include this line in the picture.) f and f^{-1} are related this way in general.

<u>Discussion</u>: Occasionally, as with our example, it may not be clear whether a graph is increasing. Maybe it will turn around offscreen and thus not be invertible after all. How do we know in general? For now we'll have to wait, but in Chapter 2 we discuss the concept of the *derivative*, which will help us with this problem.

Laboratory Exercise 1.5

The Inverse of a Function (CCH Text 1.5)

Name _____   Due Date _____

(a) Determine which of the following functions has an inverse, and explain your answer.

(b) In each case, plot a graph that supports your assertion. If you assert that a graph fails the horizontal line test, plot a horizontal line that meets the graph at least twice. (You may have to change the domain values.)

(c) If f has an inverse, find $f^{-1}(2)$ and plot the graphs of f and f^{-1} together on the same screen.

1. $f(x) = x^3 - 3x^2 + 3x$

2. $f(x) = \dfrac{x^2}{20} - 5x$

3. $f(x) = x^3 + x - 2$

4. $f(x) = x^{\frac{11}{2}} + 2x^5$

5. $f(x) = x^5(\sqrt{x} + 2)2^{-x}$

Laboratory Exercise 1.6

Restricting the Domain (CCH Text 1.5)

Name _____   Due Date _____

If we restrict the domain of a function, we can make it pass the "horizontal line test" for invertibility. For example, $f(x) = x^2$ is not invertible, but $f(x) = x^2$ restricted to $[0, \infty)$ *is* invertible, and $f^{-1}(x) = \sqrt{x}$.

1. $g(x) = x^2$ on $(-\infty, 0]$ is invertible. What is $g^{-1}(x)$?

2. Show that $h(x) = x^3 + x^2 + 1$ is not invertible.

3. Find an interval of the form (a, ∞) on which $h(x) = x^3 + x^2 + 1$ is invertible, and estimate the smallest value of a for which this is so. Explain how you arrived at your estimate.

4. Plot the graphs of $h(x)$ and $h^{-1}(x)$ on the suitably restricted interval you found in Part 3.

5. Find two other intervals on which $h(x)$ above is invertible, and plot the graphs of $h(x)$ and $h^{-1}(x)$ in these cases.

Laboratory Exercise 1.7

The Inverse of an Exponential Function (CCH Text 1.5)

Name _____   Due Date _____

1. Show that 10^x has an inverse. (Provide both graphical evidence and an algebraic argument.) For the remainder of this exercise we will refer to this inverse function as $L(x)$.

2. Find $L(10)$ and $L(20)$.

3. On the same screen graph 10^x, $L(x)$, and the line $y = x$. Explain the relationships among these three graphs.

4. Use the graph you have produced to explain why $L(x)$ is always less than x.

5. What is the domain and range of $L(x)$?

Important Information About *MATHEMATICA*'s Syntax for Logarithms

MATHEMATICA's syntax for $\log_b x$ is **Log[b,x]**. Therefore, to calculate $\log_{10} 100$, evaluating **Log[10,100]** gives the answer 2.

In *MATHEMATICA*, **Log[x]** refers to the *natural logarithm* or "log to the base e" ($e \approx 2.718$), as explained in Section 1.7 of the CCH Text. Be aware that this does not agree with the notation in the CCH Text, which uses "log" to denote the logarithm to the base 10 and "ln" to denote the logarithm to the base e.

You should also be aware that when *MATHEMATICA* simplifies any logarithmic function, it first converts to natural logarithms, and the result may not be what you expect to see. For example, if you evaluate **Log[10,x]** you will see $\dfrac{\text{Log}[x]}{\text{Log}[10]}$. It is a fact that for any three positive numbers a, $b \neq 1$, and x, $\log_b x = \dfrac{\log_a x}{\log_a b}$. *MATHEMATICA* uses this formula with e in place of a to define and calculate logarithms to bases other than e. By definition, $\log_b x = c$ means $b^c = x$. In other words, if you solve the equation $10^c = x$ for c you should get $c = \log_{10} x$, but *MATHEMATICA* will display the answer as $\dfrac{\text{Log}[x]}{\text{Log}[10]}$. Try it. Evaluate **Solve[x - 10^c==0,c]**.

Note: To enter the special number e, you must use **E**. All of *MATHEMATICA*'s constants, such as π, begin with capital letters. In *MATHEMATICA* π is **Pi**. If you just enter e, *MATHEMATICA* will treat it like any other variable, such as x or a. To verify that **E** is the correct number, approximate its value by evaluating **N[E]**. You should see 2.71828. The *MATHEMATICA* operator N gives numerical approximations of a number. Its syntax is either N[*number*] or N[*number*,n], where n is the number of decimal place accuracy you desire. Evaluate **N[E,20]** and see what you get.

> **Solved Problem 1.5: Properties of logarithms from graphs (CCH Text 1.6)**

(a) Plot the graph of $y = \log_{10}(2^x)$.

(b) Describe the graph in words and with a simple formula.

(c) Explain what this graph tells you about $\log_{10}(2^x)$.

Solution to (a): We evaluate **Clear[f]** and **f[x_] := Log[10,2^x]**, to get Out[2] in Figure 1.8. When we evaluate **Plot[f[x],{x,-1,1}]** we get the graph in Out[3].

```
In[1]:=
    Clear[f]
In[2]:=
    f[x_]:=Log[10,2^x]
In[3]:=
    Plot[f[x],{x,-1,1}]
```

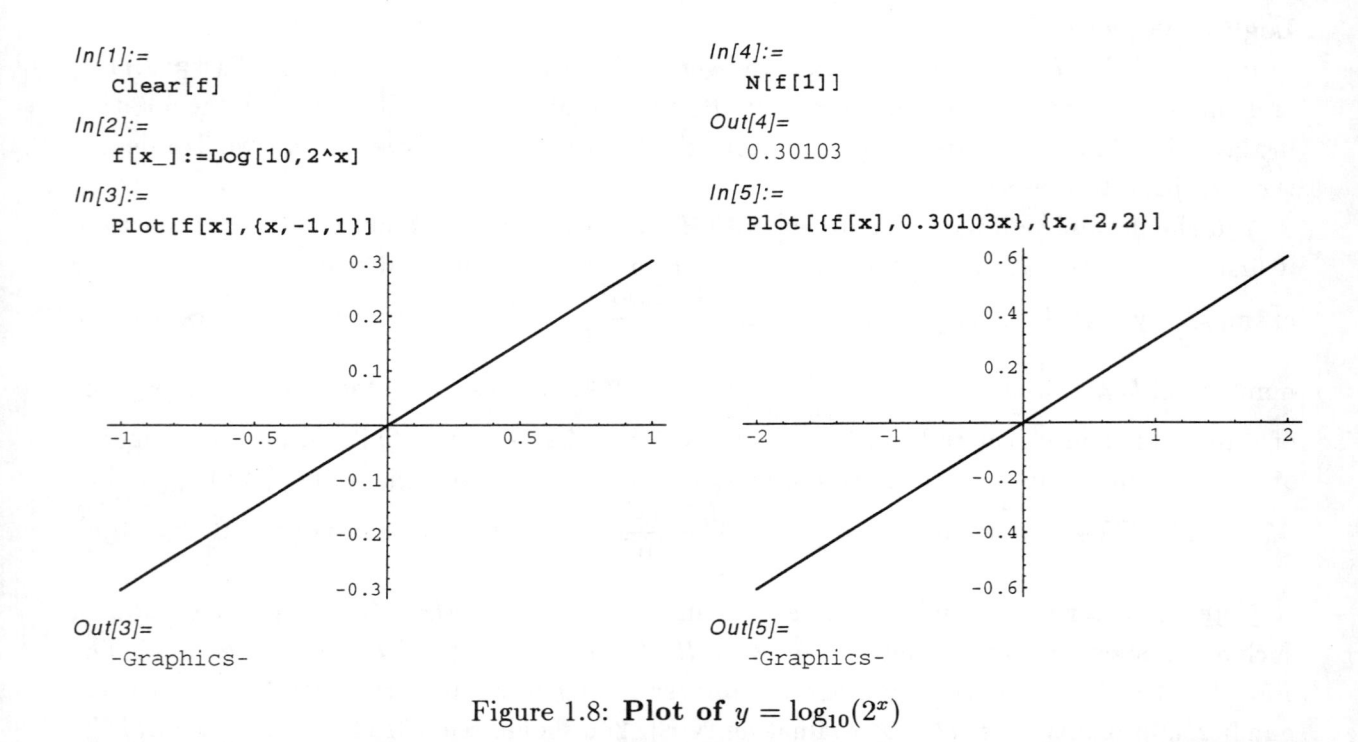

```
In[4]:=
    N[f[1]]
Out[4]=
    0.30103
In[5]:=
    Plot[{f[x],0.30103x},{x,-2,2}]
```

Out[3]=
  -Graphics-

Out[5]=
  -Graphics-

Figure 1.8: **Plot of** $y = \log_{10}(2^x)$

Solution to (b): The graph appears to be a straight line passing through the origin, very different from a log curve. (Change the domain a few times to verify this.) A straight line passing through the origin has an equation of the form $y = mx$. Therefore, there must be a number m so that $\log_{10}(2^x) = mx$. If we evaluate **N[f[1]]**, we see in Out[4] that it is approximately 0.30103. Thus the equation of the line is approximated by $y = 0.30103x$

Solution to (c): If we were to plot the linear function $y = 0.30103x$, we would see that it overlays the graph of $\log_{10}(2^x)$, verifying that $\log_{10} 2^x = 0.301029x$. As can be see from Out[5] of Figure 1.8, $\log_{10} 2 = 0.30103$ as expected.

Laboratory Exercise 1.8

Seeing Log Identities Graphically (CCH Text 1.6)

Name _____   Due Date _____

1. Plot the graph of $y = 10^{\log_{10} x}$. Describe this graph and explain what it tells you about $y = 10^{\log_{10} x}$.

2. Plot the graph of $y = 100^{\log_{10} x}$. How would you describe this graph?

3. Plot $y = x^2$ on the same screen as $y = 100^{\log_{10} x}$. When you compare the two graphs, what do you learn?

4. Discuss your observations and conclusions.

5. On the same screen plot $\log_2 x$, $\log_3 x$, $\log_4 x$, and $\log_5 x$. In general, if $b > 1$, discuss the effect of increasing b on the function $\log_b x$.

6. On the same screen graph $\log_2 x$ and $\log_{\frac{1}{2}} x$. What is the relationship between the two graphs? Conjecture how $\log_{\frac{1}{a}} x$ and $\log_a x$ are related in general. Try to prove your conjecture.

Solved Problem 1.6: Approximating the number e (CCH Text 1.7)

What is the smallest positive integer n so that $\left(1+\dfrac{1}{n}\right)^n$ approximates e to two decimal places?

Solution: We know that e is approximately 2.71828, but let's see what *MATHEMATICA* says anyway. Recall that to enter the number e we must use capital E. To get a decimal approximation, we use **N[E]**, and see the result as Out[1].

In[1]:=
   N[E]

Out[1]=
   2.71828

In[2]:=
   Clear[f]

In[3]:=
   f[n_]:=(1+(1/n))^n

In[4]:=
   f[100]

Out[4]=
   270481382942152609326719471080753083367793838278100277689020\
   10491171015143067392794394560143467445909733565137548564\
   26831251928176683242798049632232965005521797788231593800\8\
   17593329188566748424951000$1$ /
   1000000000000000000000000000000000000000000000000000000000\
   0000000000000000000000000000000000000000000000000000000000\
   0000000000000000000000000000000000000000000000000000000000\
   000000000000000000000000000

In[5]:=
   N[%]

Out[5]=
   2.70481

Figure 1.9: $\left(1+\dfrac{1}{n}\right)^n$

For an approximation to be correct to two places, it must be within ± 0.005 of e. (Look at Appendix A of the CCH Text in the part entitled *Accuracy and Error*, pages 660 to 662.) Therefore, we want $\left(1+\dfrac{1}{n}\right)^n$ to be between 2.713 and 2.723. Evaluate **Clear[f]** and **f[n_] := (1+ (1/n))^n**, which is seen as Out[3] in Figure 1.9. We might start by trying to plug in values of n and search by trial and error. Let's try 100. (Just evaluate **f[100]**.) We see the result as Out[4]. Use **N[%]** to get the approximation 2.70481 in Out[5]. That's not big enough. If we try 300, we find that $\left(1+\dfrac{1}{300}\right)^{300} \approx 2.71377$, which is between 2.713 and 2.723. So 300 works, but what is the *smallest* such n? More trial and error will be tedious and time–consuming. Is there a better way?

35

Let's try graphing. We will plot $e - (1 + 1/n)^n$ and look for the first place where the graph is below the horizontal line, $y = 0.005$.

Evaluate **g[n_]:= E - f[n]**, as seen in Out[10] in Figure 1.10. Before we plot, let's think. We want to find where the graph is below $y = 0.005$, and we know that happens out near 300. We will plot the line $y = 0.005$ on the same axes. Evaluate **Plot[{0.005,g[n]},{n,200,300}]**. The graph is seen in Figure 1.10.

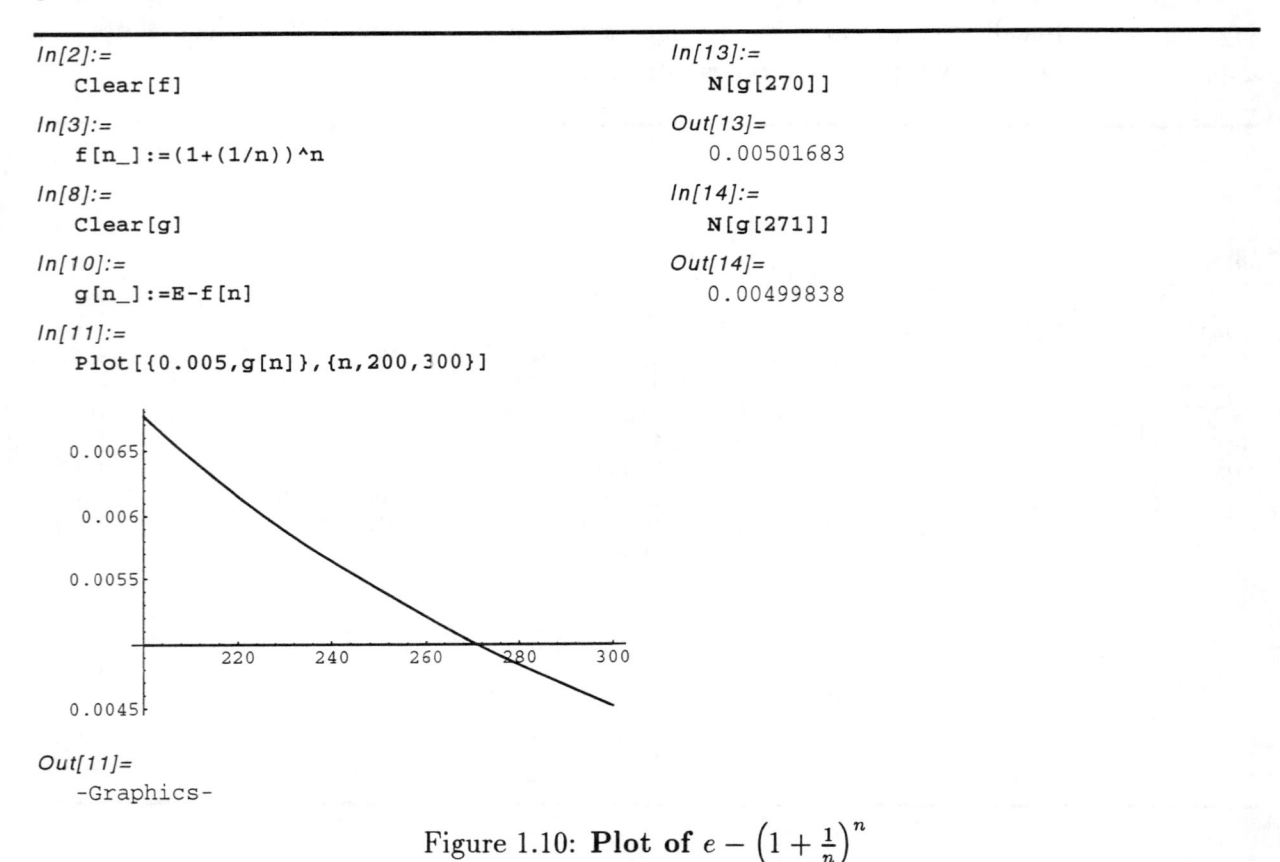

In[2]:=
  Clear[f]

In[3]:=
  f[n_]:=(1+(1/n))^n

In[8]:=
  Clear[g]

In[10]:=
  g[n_]:=E-f[n]

In[11]:=
  Plot[{0.005,g[n]},{n,200,300}]

In[13]:=
  N[g[270]]

Out[13]=
  0.00501683

In[14]:=
  N[g[271]]

Out[14]=
  0.00499838

Out[11]=
  -Graphics-

Figure 1.10: **Plot of** $e - \left(1 + \frac{1}{n}\right)^n$

Let's check it.

Evaluate **N[g[270]]**. We see that it's slightly more than 0.005, whereas **N[g[271]]** is less. The result Out[14] is 0.00499838. Therefore, the answer to the problem is 271.

<u>A Second Solution</u>: We want $e - (1 + 1/n)^n$ to be just less than 0.005. Let's try solving the equation $e - (1 + 1/n)^n = 0.005$. *MATHEMATICA* will not do it exactly, so we must find an approximate solution. The graph indicated that the function crosses the line $y = 0.005$ between 270 and 271. Now evaluate **FindRoot[0.005==g[n],{n,270}]** to see that n is approximately

270.912. Out[15] of Figure 1.11. Since n is an integer, and the graph is decreasing, we choose $n = 271$.

```
In[10]:=
    g[n_]:=E-f[n]
In[15]:=
    FindRoot[0.005-g[n]==0,{n,270}]
Out[15]=
    {n -> 270.912}
```

Figure 1.11: **Solving for** n

Laboratory Exercise 1.9

Approximating e (CCH Text 1.8)

Name _____   Due Date _____

1. Plot $\left(1+\dfrac{1}{n}\right)^n$.

2. What is the smallest positive integer n so that $\left(1+\dfrac{1}{n}\right)^n$ approximates e to three decimal places? Explain how you arrived at your answer.

Laboratory Exercise 1.10

Seeing Log Identities Graphically II (CCH Text 1.7)

Name _____   Due Date _____

1. Plot $\ln(2x) - \ln x$. How would you describe this graph?

2. What does the graph tell you about the relationship between $\ln(2x)$ and $\ln x$? Explain.

3. Plot $\ln(x^2) - 2\ln x$. How would you describe this graph?

4. What does the graph tell you about the relationship between $\ln(x^2)$ and $2\ln x$?

5. Repeat all of the above parts with "3" in place of "2."

Laboratory Exercise 1.11

Growth Rates of Functions (CCH Text 1.7)

Name _____   Due Date _____

1. On the same screen graph $\ln x$, $x^{\frac{1}{2}}$, x, x^2, and e^x.

2. What is the relationship among these 5 functions for large values of x?

3. In general, what is the relationship among logarithmic functions, power functions, and exponential functions for large values of x?

Laboratory Exercise 1.12

A Graphical Look at Borrowing Money (CCH Text 1.8)

Name _____   Due Date _____

On the same screen, plot $\left(1 + \dfrac{0.1}{n}\right)^{nt}$ for $n = 1, 2, 4, 12$; then plot $e^{0.1t}$. Explain in practical terms what the graphs tell you about borrowing money at 10% per year. Include in your explanation a discussion of the effects of time t and the number of compounding periods n.

Divide your analysis into three parts: (1) using the graphs with the domain near $t = 1$, (2) with the domain near $t = 20$, and (3) with the domain near $t = 200$.

Laboratory Exercise 1.13

Shifting and Stretching (CCH Text 1.9)

Name _____   Due Date _____

First, evaluate **Clear[f]** and **f[x_] := x^3-x**. Now evaluate the following complicated expression. Be careful that you type it in correctly.

Table[Plot[f[x]+k,{x,-2,2},DisplayFunction-> Identity],{k,0,4}]

The Table command creates a list as k varies from 0 to 5. For each k, *MATHEMATICA* plots $f(x) + k$. The option DisplayFunction ->Identity turns off the display so you do not see it. Now to see the five graphs, evaluate **Show[%,DisplayFunction -> $DisplayFunction]**. The Show command displays graphics on the same axes, and the DisplayFunction-> $DisplayFunction option turns on the display for you to see. In the problems below, you may need to change the domain.

1. Discuss in general how the graph of $f(x) + k$ varies with k.

2. Do the same procedure as above, replacing **f[x] +k** with **f[x+k]**. Discuss how the graph of $f(x + k)$ varies with k.

3. Is it true that $f(x + k) = f(x) + k$? Explain.

4. Do the same procedure, replacing **f[x]+k** with **f[x-k]**. Discuss how the graph of $f(x - k)$ varies with k.

5. How do the graphs of $f(x - k)$ compare to the case $f(x + k)$?

6. Now replace **f[x]+k** with **k*f[x]**. Discuss how the graph of $kf(x)$ varies with k.

7. Replace **f[x]+k** with **f[k*x]**. Discuss how the graph of $f(kx)$ varies with k.

8. Is it true that $f(kx) = kf(x)$? Explain.

Information About Trigonometric Functions

MATHEMATICA's syntax for the six trig functions is standard. Remember to start the function with a capital letter and use square brackets. For example, $\sin(x)$ is Sin[x] in *MATHEMATICA*. Similarly, $\arcsin x$ or $\sin^{-1}(x)$ in *MATHEMATICA* is ArcSin[x].

Trig functions always assume radian measure. Thus, Sin[30] is the sine of 30 radians, not degrees. For degrees, type "Degree" following the angle. Thus, Sin[30 Degree] is the sine of 30 degrees.

You enter the number π using Pi. Try it. Evaluate **Pi** and **N[Pi]**.

Solved Problem 1.7: Periodicity of trigonometric functions (CCH Text 1.10)

Plot $\sin^2 x$ and $\sin(x^2)$. Are they periodic? What are their periods?

Solution: Evaluate **Plot[Sin[x]^2,{x,-Pi,Pi}]** as in Out[1] of Figure 1.12.

If you enter Sin^2[x] and not Sin[x]^2, *MATHEMATICA* will be confused. *MATHEMATICA* accepts $\sin(x)^2$ but not $\sin^2 x$; whereas in textbooks, $\sin^2(x)$ and $(\sin x)^2$ are interchangeable. On the other hand, both *MATHEMATICA* and textbooks allow $\sin(x^2)$, which is not the same as $\sin(x)^2$.

Let's plot $\sin(x)^2$. We take the domain $[-2\pi, 2\pi]$. The graph appears in Out[2] of Figure 1.12 and looks periodic. The humps appear to repeat every π units, so we will conjecture that the period is π. To check this, we evaluate **Plot[Sin[x + Pi]^2,{x,-2Pi,2Pi}]** as in Out[3] Figure 1.12. The new graph is the same as the old one, and this verifies our conjecture. Notice that $\sin x$ has period 2π, but $\sin^2 x$ has period π.

Now, let's evaluate **Plot[Sin[x^2],{x,-2Pi,2Pi}]**. We see the result in Out[4] of Figure 1.12. (Notice the difference between $\sin(x^2)$ and $\sin(x)^2$. The difference is in the placement of the exponent either inside or outside of the parentheses.) As we go out the x axis, the graph seems to oscillate faster and faster, and the humps are narrower. This does *not* look periodic, and it isn't.

In[1]:=
```
Plot[Sin[x]^2,{x,-Pi,Pi}]
```

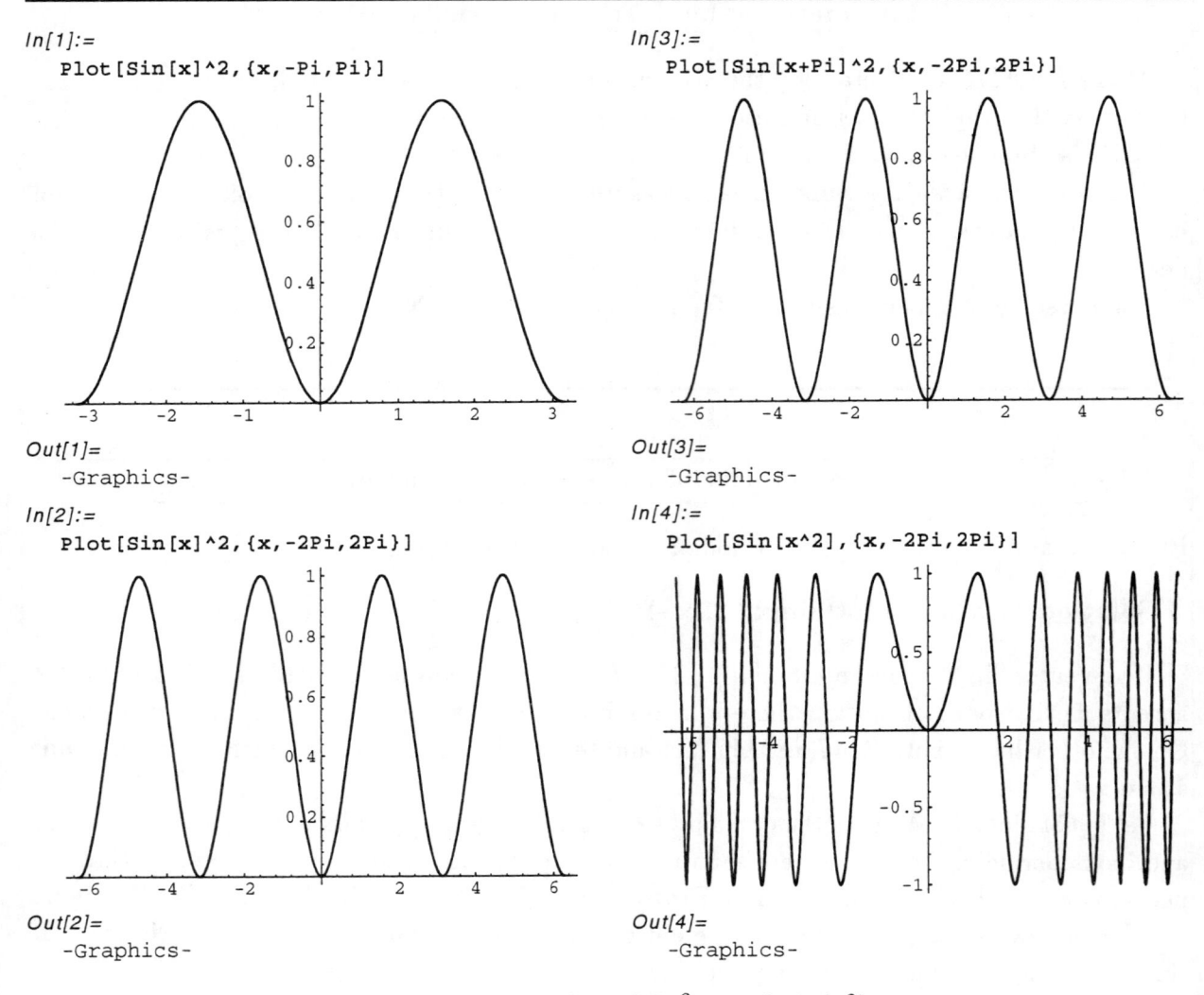

Out[1]=
-Graphics-

In[2]:=
```
Plot[Sin[x]^2,{x,-2Pi,2Pi}]
```

Out[2]=
-Graphics-

In[3]:=
```
Plot[Sin[x+Pi]^2,{x,-2Pi,2Pi}]
```

Out[3]=
-Graphics-

In[4]:=
```
Plot[Sin[x^2],{x,-2Pi,2Pi}]
```

Out[4]=
-Graphics-

Figure 1.12: **Graphs of** $\sin^2 x$ **and** $\sin(x^2)$

Laboratory Exercise 1.14

Equations Involving Trig Functions (CCH Text 1.10)

Name _____   Due Date _____

1. Plot $\cos(2x)$ and x^2.

2. Use the FindRoot command to find <u>all</u> solutions of the equation $\cos(2x) = x^2$.

3. Explain how you can be sure that you have found all solutions.

4. Use your answer in Part 2 to find all points of intersection of the graphs of $\cos(2x)$ and x^2.

Laboratory Exercise 1.15

Inverse Trig Functions (CCH Text 1.10)

Name _____   Due Date _____

1. Make plots of $\sin x$, $\arcsin x$, and $y = x$ on the same screen. Explain how the graphs are related. (Recall that in *MATHEMATICA* $\arcsin x$ is entered as ArcSin[x].)

2. Explain what you see.

3. Repeat the previous two problems for $\tan x$ and $\arctan x$.

Solved Problem 1.8: Rational functions (CCH Text 1.11)

Does $f(x) = \dfrac{2x^3 + x^2 - 15x + 7}{12x^2 - 3}$ have any horizontal asymptotes? How about $g(x) = \dfrac{2x^3 + x^2 - 15x + 7}{12x^3 - 3}$? Graph both functions together with their asymptotes.

Solution: First, we evaluate **Clear[f]** and **f[x_] := (2x^3+x^2-15x+7)/(12x^2-3)**. To test for horizontal asymptotes we look at what happens as $x \to \infty$ and $x \to -\infty$. To do this we use *MATHEMATICA*'s Limit function. Evaluate **Limit[f[x],x->Infinity]**. We get ∞, as seen in Out[3] of Figure 1.13.

Next, we evaluate **Limit[f[x],x->-Infinity]**, and we get $-\infty$ in Out[4]. These two results tell us that $f(x)$ has no horizontal asymptotes.

```
In[1]:=
   Clear[f]
In[2]:=
   f[x_]:= (2x^3+x^2-15x+7)/(12x^2-3)
In[3]:=
   Limit[f[x],x->Infinity]
Out[3]=
   Infinity
In[4]:=
   Limit[f[x],x->-Infinity]
Out[4]=
   -Infinity
```

```
In[6]:=
   Clear[g]
In[7]:=
   g[x_]:= (2x^3+x^2-15x+7)/(12x^3-3)
In[8]:=
   Limit[g[x],x->Infinity]
Out[8]=
   1
   ─
   6
In[9]:=
   Limit[g[x],x->-Infinity]
Out[9]=
   1
   ─
   6
```

Figure 1.13: **Searching for horizontal asymptotes**

When we repeat this for $g(x) = \dfrac{2x^3 + x^2 - 15x + 7}{12x^3 - 3}$, we see in outputs 8 and 9 of Figure 1.13

that both limits are $\frac{1}{6}$, indicating a horizontal asymptote, $y = \frac{1}{6}$.

In Figure 1.14 we have graphed both functions (with $g(x)$ thickened) and added the horizontal asymptote, $y = \frac{1}{6}$, for the second function.

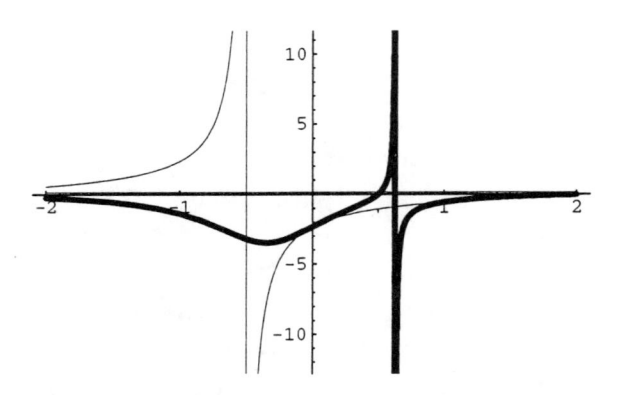

Figure 1.14: **Horizontal asymptotes displayed**

Solved Problem 1.9: Vertical asymptotes (CCH Text 1.11)

Let $f(x) = \dfrac{2x^3 + x^2 - 15x + 7}{12x^2 - 3}$. Plot $f(x)$. How many vertical asymptotes do you see? How many do you expect to see? Explain. Add the graphs of the vertical asymptotes to the picture.

Solution: We evaluate **Clear[f]** and **f[x_] := (2x^3+x^2-15x+7)/(12x^2-3)**, which appears in Out[2] of Figure 1.15. When we evaluate **Plot[f[x],{x,-2,2}]**, we see the graph in Out[3]. From this it appears there is a vertical asymptote between $x = 0$ and $x = -1$.

```
In[1]:=
   Clear[f]
In[2]:=
   f[x_]:=(2x^3+x^2-15x+7)/(12x^2-3)
In[3]:=
   Plot[f[x],{x,-2,2}]
```

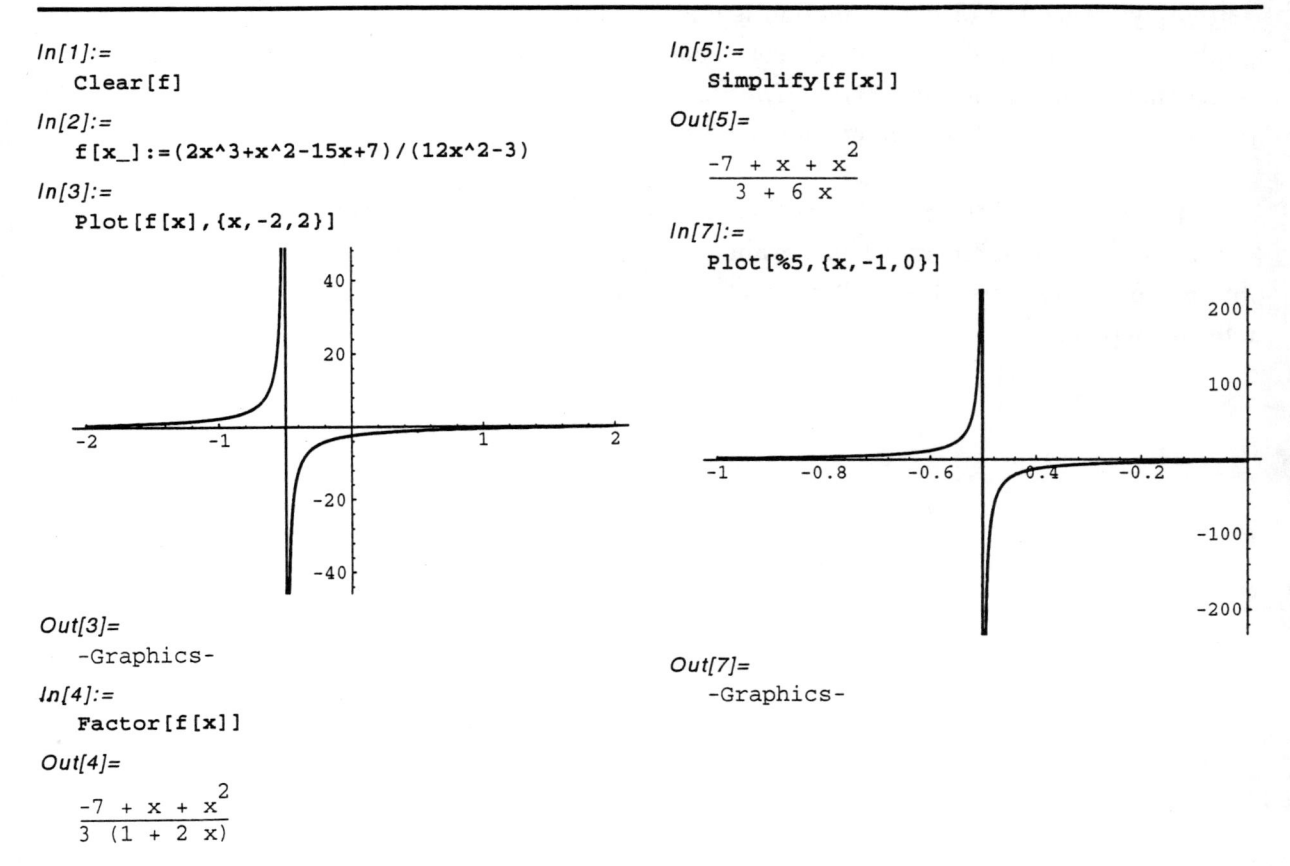

Out[3]=
   -Graphics-

```
In[4]:=
   Factor[f[x]]
```
Out[4]=
$$\frac{-7 + x + x^2}{3(1 + 2x)}$$

```
In[5]:=
   Simplify[f[x]]
```
Out[5]=
$$\frac{-7 + x + x^2}{3 + 6x}$$

```
In[7]:=
   Plot[%5,{x,-1,0}]
```

Out[7]=
   -Graphics-

Figure 1.15: **A phantom asymptote**

To test for vertical asymptotes we look for values of x where the denominator is zero. $12x^2 - 3 = 0$ for $x = \pm\frac{1}{2}$. Therefore, we are not surprised to see $x = -\frac{1}{2}$, which must be the asymptote in the graph above. But where is the one at $x = \frac{1}{2}$? To see why it is missing, we will factor the numerator and denominator.

We use *MATHEMATICA*'s Factor function. We evaluate **Factor[f[x]]** and we see Out[4] of Figure 1.15.

We can now see why there is no vertical asymptote at $x = \frac{1}{2}$. There is a common factor of $2x - 1$ in both the numerator and denominator. When they are canceled, we see that only the $2x + 1$ remains in the denominator. Hence, there is just one asymptote. We could also evaluate **Simplify[f[x]]**, which yields Out[5]. Plotting it gives the same graph as before, and now we expect only the single asymptote.

Remark: Since the original function is undefined at $x = \frac{1}{2}$, we would expect a "hole" in the graph. If we put the domain as $(0.4999, 0.5001)$, *MATHEMATICA* gives error messages indicating that it has encountered 1/0. *MATHEMATICA* will plot the other values it can compute and will give a plot where you cannot see the gap.

A trick is required to graph the vertical line $x = -\frac{1}{2}$. The graph is the collection of all points $(-\frac{1}{2}, y)$, where y is arbitrary. Thus, evaluating **p=ParametricPlot[{-1/2,y},{y,-20,20}]**, we obtain the vertical line. Then **p2=Plot[f[x],{x,-2,2}]** plots $f(x)$. To put them together, evaluate **Show[p,p2]**.

Laboratory Exercise 1.16

Asymptotes (CCH Text 1.11)

Name _____   Due Date _____

1. Make plots of the two functions in Solved Problem 1.8 that clearly show their "end" behavior, that is, the horizontal asymptotes or lack of them.

2. Explain how the graphs illustrate the results obtained in Solved Problem 1.8.

3. Find the vertical asymptotes of the second function in Solved Problem 1.8. Explain how you arrived at your answers.

4. Let $g(x) = \dfrac{3x^3 - 17x^2 + 16x - 4}{3x^3 - 2x^2 + 3x - 2}$. Find all the vertical asymptotes of $g(x)$.

5. Find all the horizontal asymptotes of $g(x)$.

6. Plot $g(x)$ so the graph shows the asymptotes. (More than one plot may be necessary.)

7. Let $h(x) = \dfrac{3x^3 - 17x^2 + 16x - 4}{3x^3 - 2x^2 - 3x + 2}$. Find all the vertical asymptotes of $h(x)$.

8. Find all the horizontal asymptotes of $h(x)$.

9. Plot $h(x)$ so the graph shows the asymptotes. (More than one plot may be necessary.)

Chapter 2
Key Concept: The Derivative

In this chapter we investigate how average velocity over a time interval can be used to define and calculate instantaneous velocity. This idea is extended to explore average and instantaneous rates of change for general functions. As usual, graphing will be employed extensively, and you are advised to consult Sections 3 and 4 of Appendix I for details on plotting with *MATHEMATICA*.

Solved Problem 2.1: Calculating velocities (CCH Text 2.1)

Taking into account air resistance, a suitcase dropped from an airplane falls $968(e^{-0.18t}-1)+176t$ feet in t seconds.

(a) Find the average velocity of the suitcase over the time intervals $t = 1.99$ to $t = 2$ and $t = 2$ to $t = 2.01$. Use this to estimate the instantaneous velocity of the suitcase at time $t = 2$.

(b) Plot the graph of the average velocity from $t = 2$ to $t = 2 + h$ as a function of h and use it to estimate the instantaneous velocity of the suitcase at time $t = 2$.

(c) Find the instantaneous velocity of the suitcase at time $t = 2$ by calculating the appropriate limit. Compare this to the two estimates in Parts (a) and (b).

<u>Solution to (a)</u>: First evaluate **Clear[s]** and **s[t_] := 968(E^(-0.18t)-1)+176t**, as seen in In[2] of Figure 2.1. Be sure to use ":=" rather than "=". This tells *MATHEMATICA* that we want a *definition* rather than an equation. (See Section 5 of Appendix I.) The average velocity over the time interval [1.99, 2] is the distance traveled divided by the elapsed time:

$$\frac{s(2) - s(1.99)}{0.01}.$$

When we evaluate **(s[2] - s[1.99])/0.01**, we obtain the average velocity, 54.3274 feet per second, in Out[3]. Similarly, the average velocity over the time interval [2, 2.01] is $\frac{s(2.01) - s(2)}{0.01} \approx$ 54.5462 feet per second. Either of these is a reasonable estimate for the instantaneous velocity, as is their average, 54.4368 feet per second. The CCH Text customarily uses $\frac{s(2.01) - s(2)}{0.01} \approx$ 54.5462 feet per second.

```
In[1]:=
  Clear[s]
In[2]:=
  s[t_]:=968(E^(-0.18t)-1)+176t
In[3]:=
  (s[2]-s[1.99])/0.01
Out[3]=
  54.3274
In[4]:=
  (s[2.01]-s[2])/0.01
Out[4]=
  54.5462

In[5]:=
  Plot[(s[2+h]-s[2])/h,{h,-0.01,0.01}]
```

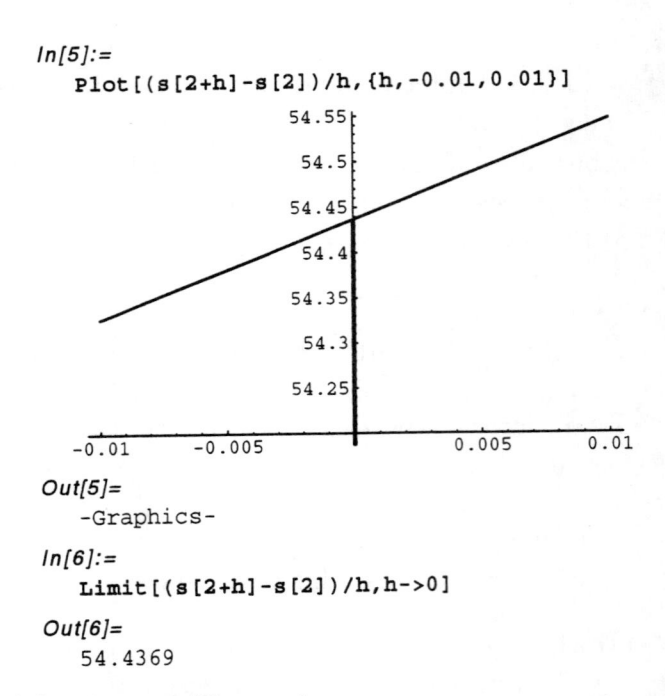

[Graph showing linear plot from approximately 54.25 to 54.55, crossing y-axis near 54.4]

```
Out[5]=
  -Graphics-
In[6]:=
  Limit[(s[2+h]-s[2])/h,h->0]
Out[6]=
  54.4369
```

Figure 2.1: **Average velocities for a falling suitcase**

Solution to (b): The average velocity over the time interval $t = 2$ to $t = 2 + h$ is $\dfrac{s(2+h) - s(2)}{h}$. We evaluate **Plot[(s[2+h]-s[2])/h,{h,-0.01,0.01}]** to get the picture in Out[5].

We are interested in the value of $\dfrac{s(2+h) - s(2)}{h}$ when h is near 0. Place the mouse pointer in the graphics window, press the button, hold the Crtl key on a PC or the Command-Apple keys on a Macintosh, and move the mouse pointer to the where the graph crosses the y axis. The y coordinate is given on the left side of the bottom border of the *MATHEMATICA* window. We see that the y value here is approximately 54.4. This should be a reasonable estimate for the instantaneous velocity. Remark: There is actually a hole in this graph at $h = 0$. Look at the definition of the function and explain why.

Solution to (c): The exact value of the instantaneous velocity is $\lim\limits_{h \to 0} \dfrac{s(2+h) - s(2)}{h}$. We evaluate **Limit[(s[2+h] -s[2])/h,h-> 0]**. The result is seen in Out[6] of Figure 2.1 and we obtain the answer, 54.4369. Our estimate in (a) was better than that in (b), which could be improved by using a smaller interval about h as the domain.

Making Tables of Values

The Table function in *MATHEMATICA* can be used to make a table of values. For example, let's make a table of average velocities for Solved Problem 2.1 over the intervals

$$[1.95, 2], [1.96, 2], \ldots [2, 2.04], [2, 2.05]$$

Table[{h, (s(2+h)-s(2))/h},{ h, -0.05, 0.05, 0.01}]. This will display the pairs $[h, \frac{s(2+h) - s(2)}{h}]$ as the variable h ranges from -0.05 to 0.05 in steps of 0.01. We show the table of average velocities in Out[1] of Figure 2.2. Notice that *MATHEMATICA* put $\{3.4694510^{-18}, 0.\}$ instead of the pair $\{h, (s(2+h) - s(2))/h\}$ for $h = 0$ (or $\{0, Indeterminate\}$ on a Macintosh). Why did it do that?

```
In[2]:=
    s[t_]:=968(E^(-0.18t)-1)+176t
In[3]:=
    Table[{h,(s[2+h]-s[2])/h},{h,-0.05,0.05,0.01}]
                                            1
    Power::infy: Infinite expression  --  encountered.
                                            0.

    Infinity::indet:
         Indeterminate expression 0. ComplexInfinity encountered.
Out[3]=
    {{-0.05, 53.8882}, {-0.04, 53.9982}, {-0.03, 54.1081},
     {-0.02, 54.2178}, {-0.01, 54.3274}, {0., Indeterminate},
     {0.01, 54.5462}, {0.02, 54.6554}, {0.03, 54.7645},
     {0.04, 54.8735}, {0.05, 54.9823}}
```

Figure 2.2: **A table of average velocities**

Laboratory Exercise 2.1

Average Velocity (CCH Text 2.1)

Name _____   Due Date _____

1. You drive for one minute at a constant velocity of 60 miles per hour. You then instantly slow down and drive for one more minute at a constant velocity of 40 miles per hour. Assuming you lost no time in slowing down, what is your average velocity for the two minutes?

2. You drive for one mile at a constant velocity of 60 miles per hour. You then instantly slow down and drive for one more mile at a constant velocity of 40 miles per hour. Assuming you lost no time in slowing down, what is your average velocity for the two miles? (<u>Note</u>: 50 miles per hour is not the right answer. If that is what you got, think once more about the definition of average velocity.)

3. You drive for one minute at a constant velocity of 30 miles per hour. You then want to instantly speed up and drive another minute so that your average velocity for the two minutes is 60 miles per hour. Assuming you lose no time in speeding up, what velocity must you drive for the second minute?

4. You drive for one mile at a constant velocity of 30 miles per hour. You then want to instantly speed up and drive another mile so that your average velocity for the two miles is 60 miles per hour. Assuming you lose no time in speeding up, what velocity must you drive for the second mile? (Be careful.)

Laboratory Exercise 2.2

Falling with a Parachute (CCH Text 2.1)

Name _____   Due Date _____

A person jumping from an airplane with a parachute, falls $s(t) = 12.5(e^{-1.6t} - 1) + 20t$ feet in t seconds. Answer the following questions about the parachutist. (Be sure to present your answers with appropriate units.)

1. Find the parachutist's average velocity over the time intervals $[0.99, 1]$ and $[1, 1.01]$. Estimate the instantaneous velocity at $t = 1$. Explain how you got your answers.

2. Plot the graph of the parachutist's average velocity from $t = 1$ to $t = 1 + h$ as a function of h. Use this graph to estimate the instantaneous velocity at $t = 1$ and compare your answer with the one you got in Part 1.

3. Calculate the parachutist's instantaneous velocity at $t = 1$ by using *MATHEMATICA* to find the appropriate limit.

4. Plot the graph of $s(t)$. Use this graph to describe the *velocity* of the parachutist as a function of time. Include a discussion of what happens for "large" values of t.

Solved Problem 2.2: Average rates of change (CCH Text 2.2)

Let $f(x) = \dfrac{4x}{4x^2 + 1}$.

(a) Find the average rate of change of $f(x)$ from $x = 0.25$ to $x = 0.55$.

(b) Find the equation of the corresponding secant line.

(c) Plot the graphs of $f(x)$ and of the secant line.

(d) Repeat all the above steps for $x = 0.25$ to $x = 0.35$. Explain what you observe.

<u>Solution to (a)</u>: We first evaluate **Clear[f]** and **f[x_] := 4x/(4x^2+1)**. The average rate of change of f from 0.25 to 0.55 is $\dfrac{f(0.55) - f(0.25)}{0.3}$. Thus we evaluate **(f[0.55]-f[0.25])/0.3** to get the answer 0.651584 in Out[3] of Figure 2.3.

<u>Solution to (b)</u>: The secant line is the line passing through the two points $(0.25, f(0.25))$ and $(0.55, f(0.55))$. Its slope is the average rate of change we found in Part (a). Thus the equation of this line is

$$y = 0.651584(x - 0.25) + f(0.25)$$

We evaluate **Clear[y]** and **y[x_] = 0.651584(x-0.25) + f[0.25]** to obtain the equation in In[5] of Figure 2.3.

<u>Solution to (c)</u>: We evaluate **Plot[{f[x],y[x]}, {x,0.1,0.8},AxesOrigin->{0.1,0.4}]**. These graphs appear in Out[6] of Figure 2.3. We used the *MATHEMATICA* graphics option AxesOrigin ->{c,d} to get the displayed axes in a convenient location. See what you get if you omit that option.

<u>Solution to (d)</u>: The average rate of change over the time interval $[0.25, 0.35]$ is

$$\dfrac{f(0.35) - f(0.25)}{0.1} = 1.39597$$

and the equation of the secant line is

$$y = 1.39597(x - 0.25) + f(0.25) = 1.39597x + 0.451007$$

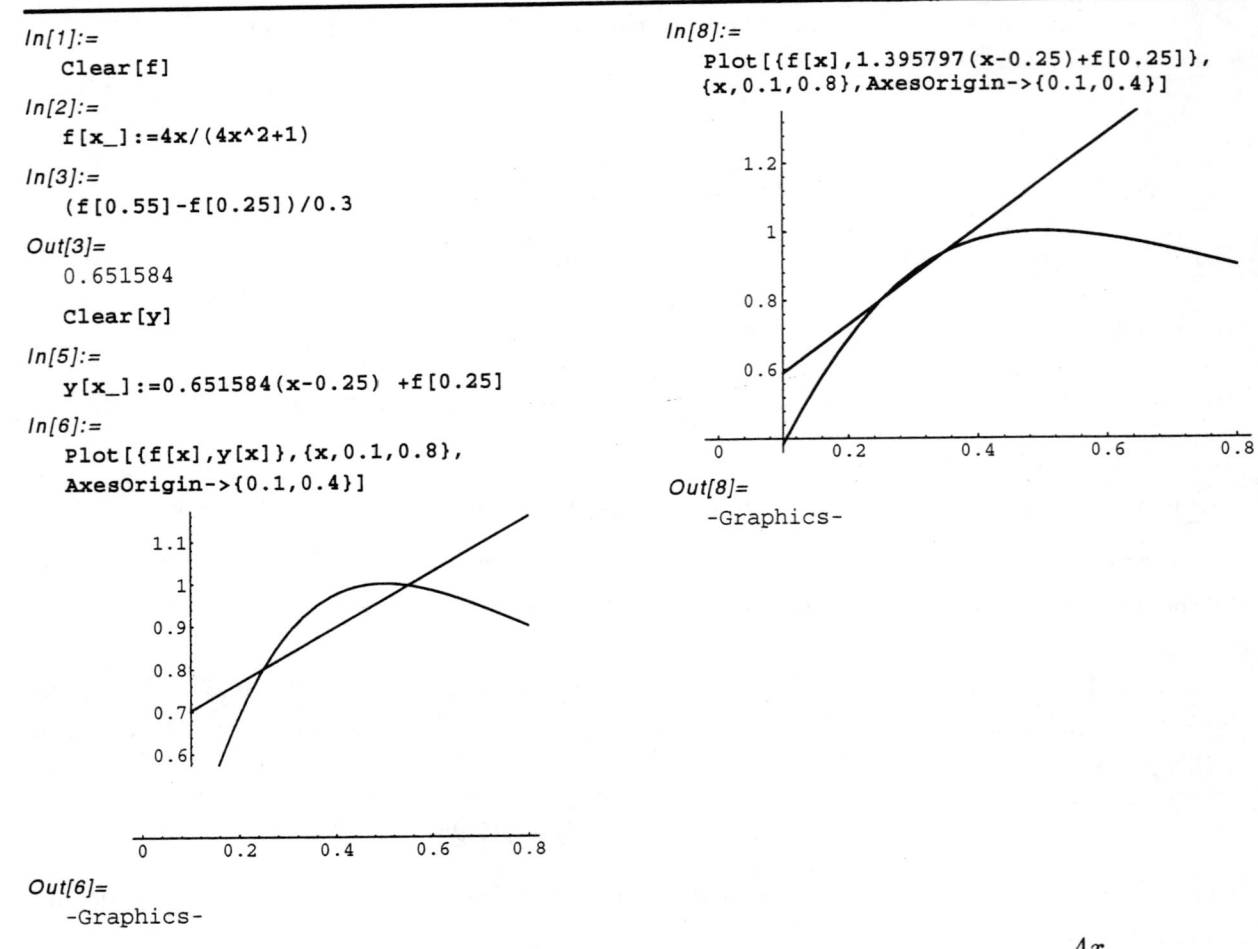

```
In[1]:=
  Clear[f]
In[2]:=
  f[x_]:=4x/(4x^2+1)
In[3]:=
  (f[0.55]-f[0.25])/0.3
Out[3]=
  0.651584

  Clear[y]
In[5]:=
  y[x_]:=0.651584(x-0.25) +f[0.25]
In[6]:=
  Plot[{f[x],y[x]},{x,0.1,0.8},
  AxesOrigin->{0.1,0.4}]
```

```
In[8]:=
  Plot[{f[x],1.395797(x-0.25)+f[0.25]},
  {x,0.1,0.8},AxesOrigin->{0.1,0.4}]
```

Out[8]=
-Graphics-

Out[6]=
-Graphics-

Figure 2.3: **Secant lines and average rates of change of** $f(x) = \dfrac{4x}{4x^2+1}$

The graphs of the function f and both secant lines appear in Out[8] of Figure 2.3.

The value 0.35 is much closer to 0.25 ($h = 0.1$) than is 0.55 ($h = 0.2$), and we observe that the secant line is now nearly *tangent* to the curve at $(0.25, f(0.25))$. The reader is encouraged to plot secant lines for still smaller values of h (and for negative values as well — to observe convergence to tangency.

Solved Problem 2.3: Instantaneous rates of change (CCH Text 2.2)

Let $f(x) = \dfrac{4x}{1+4x^2}$.

(a) Find the instantaneous rate of change of f at $x = 0.25$ using the definition.

(b) Find the equation of the corresponding tangent line.

(c) Plot the graphs of f and of the tangent line. Zoom in on the point of tangency and explain what you see.

(d) Plot the graph of the difference quotient from Part (a). Explain how you can use this graph to estimate the derivative of f at 0.25.

Solution to (a): As in Solved Problem 2.2, we evaluate **Clear[f]** and **f[x_] := (4x)/(4x^2+1)**. The instantaneous rate of change (or derivative) of f at 0.25 is $\lim_{h \to 0} \dfrac{f(0.25 + h) - f(0.25)}{h}$. Thus evaluating **Limit[(f[0.25+h]-f[0.25])/h,h->0]**, we obtain 1.92 as seen in Out[3] of Figure 2.4.

In[1]:=
  Clear[f]

In[2]:=
  f[x_]:=4x/(4x^2+1)

In[3]:=
  Limit[(f[0.25 +h]-f[0.25])/h,h->0]

Out[3]=
  1.92

In[4]:=
  Clear[y]

In[5]:=
  y[x_]:=1.92(x-0.25)+f[0.25]

In[6]:=
  Plot[{f[x],y[x]},{x,0.2,0.3}]

Out[6]=
  -Graphics-

In[7]:=
  Plot[{f[x],y[x]},{x,0.24,0.26}]

Out[7]=
  -Graphics-

In[9]:=
  Plot[{f[x],y[x]},{x,0.249,0.251}]

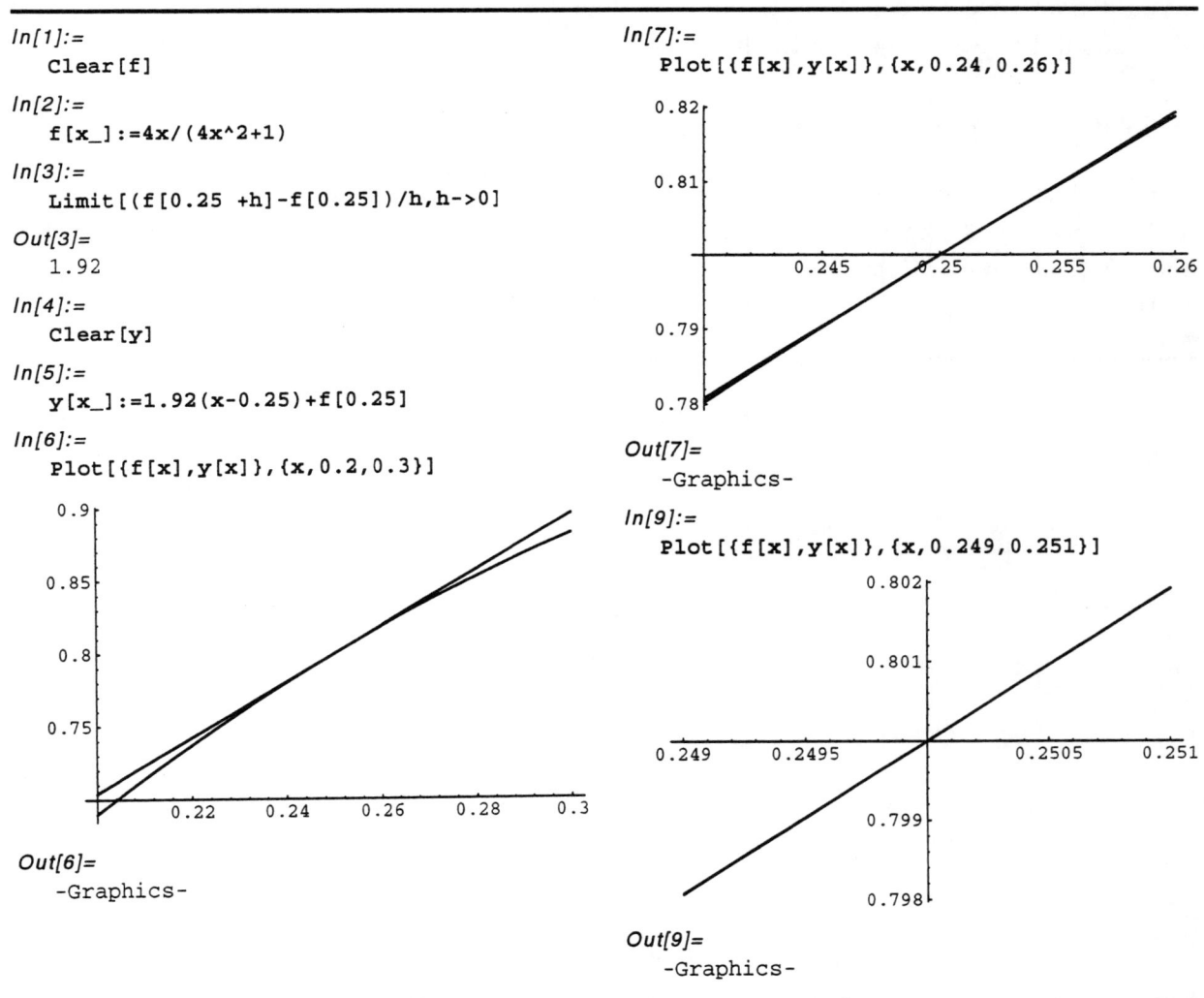

Out[9]=
  -Graphics-

Figure 2.4: **A tangent line to** $f(x) = \dfrac{4x}{4x^2 + 1}$

Solution to (b): The tangent line is the line through the point $(0.25, f(0.25))$ with slope equal to the instantaneous rate of change that we found in Part (a). So the equation of this line is $y = 1.92(x - 0.25) + f(0.25)$. We evaluate **Clear[y]** and **y[x_]:= 1.92(x-0.25) + f[0.25]** to obtain the equation in In[5].

Solution to (c): Evaluate **Plot[{f[x],y[x]},{x,0.2,0.3}]** to get Out[6] of Figure 2.4. You should compare this picture with the graphs of the secant lines in Solved Problem 2.2.

We change to domain to zoom in on $x = 0.25$. Evaluate **Plot[{f[x],y[x]},{x,0.24,0.26}]**, and then **Plot[{f[x],y[x]},{x,0.249,0.251}]**. The resulting graphs are in Out[7] and Out[9] of Figure 2.4.

We observe that the graphs of f and its tangent line grow closer together and seem to be actually merging. This is a demonstration of the principal of *local linearity*. That is, if a function has a derivative at a point, then the function looks and behaves very much like its tangent line if we zoom in close enough. In the last output, we can scarcely distinguish the two graphs.

Solution to (d): Since the derivative of f at 0.25 is $\lim_{h \to 0} \dfrac{f(0.25 + h) - f(0.25)}{h}$, we are interested in what happens to the graph of $\dfrac{f(0.25 + h) - f(0.25)}{h}$ when h is near 0.

We can estimate the derivative by looking at the y coordinate of the point where the graph of the difference quotient appears to cross the y axis. Evaluate **Plot[(f[0.25+h]-f[0.25])/h,{h,-0.01,0.01}]**. The graph appears in Out[10] of Figure 2.5.

```
In[2]:=
    f[x_]:=4x/(4x^2+1)
In[10]:=
    Plot[(f[0.25+h]-f[0.25])/h,{h,-0.01,0.01}]
```

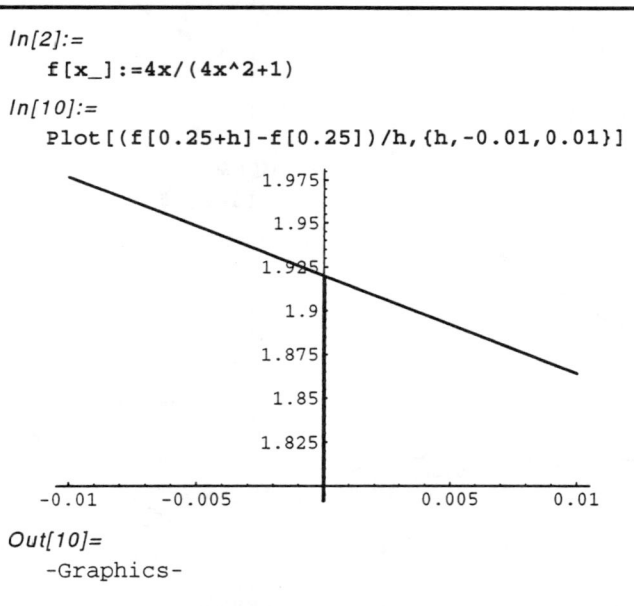

Out[10]=
-Graphics-

Figure 2.5: **Plot of the difference quotient**

We see that the graph crosses the y axis just below 1.925 Thus, the y value is very close to the actual derivative, 1.92, that we computed earlier. By taking domain values closer to $h = 0$ we could obtain a more accurate estimate.

It should be emphasized that the difference quotient $\dfrac{f(0.25 + h) - f(0.25)}{h}$ is not defined at $h = 0$, so there is actually a hole in the graph there. What we are estimating is the *limit* of this difference quotient, not its value at $h = 0$.

Laboratory Exercise 2.3

Slopes and Average Rates of Change (CCH Text 2.2)

Name _____   Due Date _____

Let $f(x) = 3x - 2x^2$.

1. Find the average rate of change of f from $x = 0.5$ to $x = 0.9$.

2. Find the equation of the corresponding secant line.

3. Plot the graphs of f and the secant line.

4. Repeat Parts 1, 2, and 3 for $x = 0.5$ to $x = 0.51$. Explain what you observe.

5. Zoom in on the point $(0.5, f(0.5))$ taking a small domain interval around 0.5. Show your plot, and explain what you observe about the two graphs in Part 4.

6. Now reconsider the graph of $3x - 2x^2$. By making a small domain interval, zoom in on the point $(0.5, f(0.5))$ until the graph looks like a straight line. Show your plot and explain how you can use this graph to estimate the slope of this "line."

Laboratory Exercise 2.4

Tangent Lines and Rates of Change (CCH Text 2.2)

Name _____   Due Date _____

Let $f(x) = 3x - 2x^2$.

1. Find the instantaneous rate of change of f at $x = 0.5$ *using the definition.*

2. Find the equation of the corresponding tangent line.

3. Plot the graphs of f and of the tangent line on the same screen. Zoom in on the point $(0.5, f(0.5))$ until the two graphs are indistinguishable. How small must the domain interval be for this to happen?

4. Plot the graph of $\dfrac{f(0.5 + h) - f(0.5)}{h}$. Explain how you can use this graph to estimate the derivative of f at 0.5.

Laboratory Exercise 2.5

The Derivative of the Gamma Function (CCH Text 2.2)

Name _____   Due Date _____

MATHEMATICA includes a function called *the Gamma function*. Evaluating **Gamma[x]** returns the value of the Gamma function at x. This function is very important in mathematics, and you will encounter it later when you get to the subject of improper integrals. (Basically, the Gamma function fills in the gaps in the graph of $(n-1)!$.)

1. **Simplify** $\Gamma(1)$, $\Gamma(\frac{1}{2})$, and $\Gamma(3)$. What do you get?

2. Try to find the derivative of $\Gamma(x)$ at $x = 1$ by asking *MATHEMATICA* to calculate the appropriate limit and by using *MATHEMATICA*'s derivative function D[ ]. This is accomplished by evaluating **D[Gamma[x],x]/.x->1**. The function D takes the derivative of the first argument with respect the variable the second argument. When you evaluate an expression using /.x->1, *MATHEMATICA* replaces each occurrence of x with 1. You can get a numerical approximation using N[ ].

   In this laboratory, we calculate $\Gamma'(1)$ using difference quotients.

3. Estimate the derivative by approximating $\dfrac{\Gamma(1+h) - \Gamma(1)}{h}$ with $h = 0.001$.

4. Plot the graph of $\Gamma(x)$.

5. Zoom in on the point $(1, \Gamma(1))$ until the graph looks like a straight line. What domain is required to accomplish this? Show your plot and use this graph to estimate the slope of this "line." Explain all your steps clearly, and compare your answer with that obtained in Part 3.

6. Plot the graph of $\dfrac{\Gamma(1+h) - \Gamma(1)}{h}$. Use this graph to estimate the derivative of Γ at 1. Explain what you did, and compare your answer with those obtained in Parts 3 and 5.

7. Find the equation of the tangent line to $\Gamma(x)$ at $x = 1$ using the estimate for the slope from Part 6.

8. Plot the graphs of $\Gamma(x)$ and of the tangent line. Zoom in on the point of tangency and explain how these graphs support your estimate of $\Gamma'(1)$.

Solved Problem 2.4: The derivative function (CCH Text 2.3)

Let $f(x) = \dfrac{x}{1+x^2}$.

(a) Use *MATHEMATICA* to calculate $f'(x)$ in two ways: (i) directly and (ii) by the definition.

(b) On the same axes plot the graphs of f and f'. Describe the relationship between the sign of f' and the graph of f.

Solution to (a): We first evaluate **Clear[f,df]** and **f[x_] := x/(1+x^2)**. To calculate the derivative directly, evaluate **D[f[x],x]**. The result is Out[3] in Figure 2.6. To create the derivative function, evaluate either **df[x_]:= %** or **df[x_] := (-2x^2)/(1+x^2)^2 + (1+x^2)^(-1)**.

In[1]:=
   Clear[f,df]

In[2]:=
   f[x_]:=x/(1+x^2)

In[3]:=
   D[f[x],x]

Out[3]=

$$\dfrac{-2x^2}{(1+x^2)^2} + \dfrac{1}{1+x^2}$$

In[7]:=
   df[x_]:=%3

In[8]:=
   df[x]

Out[8]=

$$\dfrac{-2x^2}{(1+x^2)^2} + \dfrac{1}{1+x^2}$$

In[9]:=
   Simplify[df[x]]

Out[9]=

$$\dfrac{1-x^2}{(1+x^2)^2}$$

In[10]:=
   Limit[(f[x+h]-f[x])/h,h->0]

Out[10]=

$$\dfrac{-2x^2}{(1+x^2)^2} + \dfrac{1}{1+x^2}$$

In[11]:=
   Simplify[%]

Out[11]=

$$\dfrac{1-x^2}{(1+x^2)^2}$$

In[13]:=
   Plot[{f[x],df[x]},{x,-2,2},PlotStyle -> {Thickness[0.001],Thickness[0.008]}]

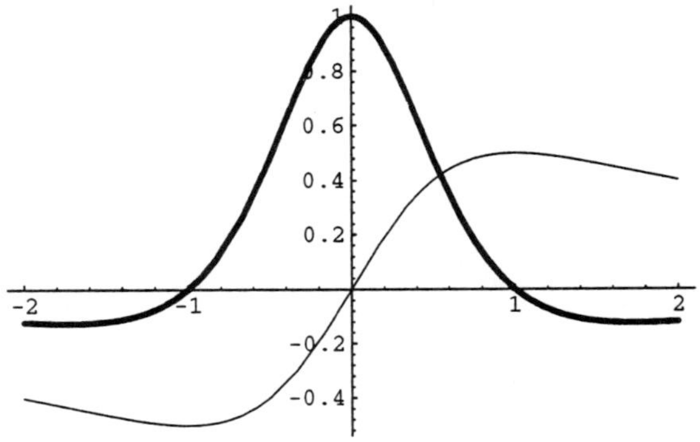

Out[13]=
   -Graphics-

Figure 2.6: $\dfrac{x}{x^2+1}$ and its derivative

83

Evaluate **Simplify[df[x]]** to obtain the derivative in Out[9].

To calculate the derivative using the definition, we evaluate **Limit[(f[x+h]-f[x])/h,h->0]**. and obtain Out[10] of Figure 2.6. Use Simplify[%] to get the derivative displayed in Out[11]. (Note that the outputs 9 and 11 are identical, as we expect.)

Solution to (b): Evaluate **Plot[{f[x],df[x]}, {x,-2,2}, PlotStyle -> {Thickness[.001], Thickness[.008]}]**. The Thickness option allows one to easily distinguish between the function $f(x)$ (the lighter curve) and its derivative (the darker curve). If you have a color monitor you could use the RGBColor command described in Section 3 of Appendix I.

The graphs are displayed in Out[13] of Figure 2.6. We observe that the graph of f is increasing where the graph of f' is above the x axis (on the interval $(-1, 1)$) and that the graph of f is decreasing where the graph of f' is below the x axis (on the intervals $(-\infty, -1)$ and $(1, \infty)$).

Laboratory Exercise 2.6

The Meaning of the Sign of f' (CCH Text 2.3)

Name _____   Due Date _____

This is a graphical demonstration to highlight the following important principle: If f' is positive then f is increasing. If f' is negative then f is decreasing. This demonstration requires a color monitor.

1. First evaluate **Clear[f,df]** and **f[x_]:= Sin[x]**. Create the derivative function by evaluating **df[x_]:= D[f[t],t]/.t->x**. Next evaluate the following *exactly* as it appears:

    ```
    Plot[f[x],df[x],x,-2Pi,2Pi,PlotStyle->RGBColor[1,0,0],RGBColor[0,1,0]]
    ```

    Explain what you see. Which color represents $\sin(x)$? What does the green curve being below the x axis signify? What does the green curve crossing the x axis signify?

2. Repeat the experiment above for $x^3 - x$. Use the domain $[-2, 2]$.

Laboratory Exercise 2.7

Recovering f from f' (CCH Text 2.5)

Name _____   Due Date _____

Suppose f is a function with the property that $f(0) = 0$ and $f'(x) = \ln(x^4 - x^2 + 0.5)$

1. Plot the graph of f' and determine the intervals where it is positive and where it is negative.

2. Use *MATHEMATICA* to calculate f''.

3. Plot the graph of f'' and determine the intervals where it is positive and where it is negative.

4. Use the information from Parts 1 and 3 to sketch the graph of f. Your graph should accurately reflect this information.

Solved Problem 2.5: Calculating limits graphically (CCH Text 2.7)

Use graphs to estimate the value of the two limits $\lim_{x \to 0} (1 + 3x)^{\frac{1}{x}}$ and $\lim_{x \to 0} \sin\left(\frac{1}{x}\right)$. Use *MATHEMATICA* to calculate the limits, and compare them with your estimates.

<u>Solution</u>: The first step is to evaluate **Plot[(1+3x)^(1/x),{x,-.1,.1}]**. Out[1] of Figure 2.7 indicates trouble at $x = 0$.

```
Plot[(1+3x)^(1/x),{x,-.1,.1}]
```

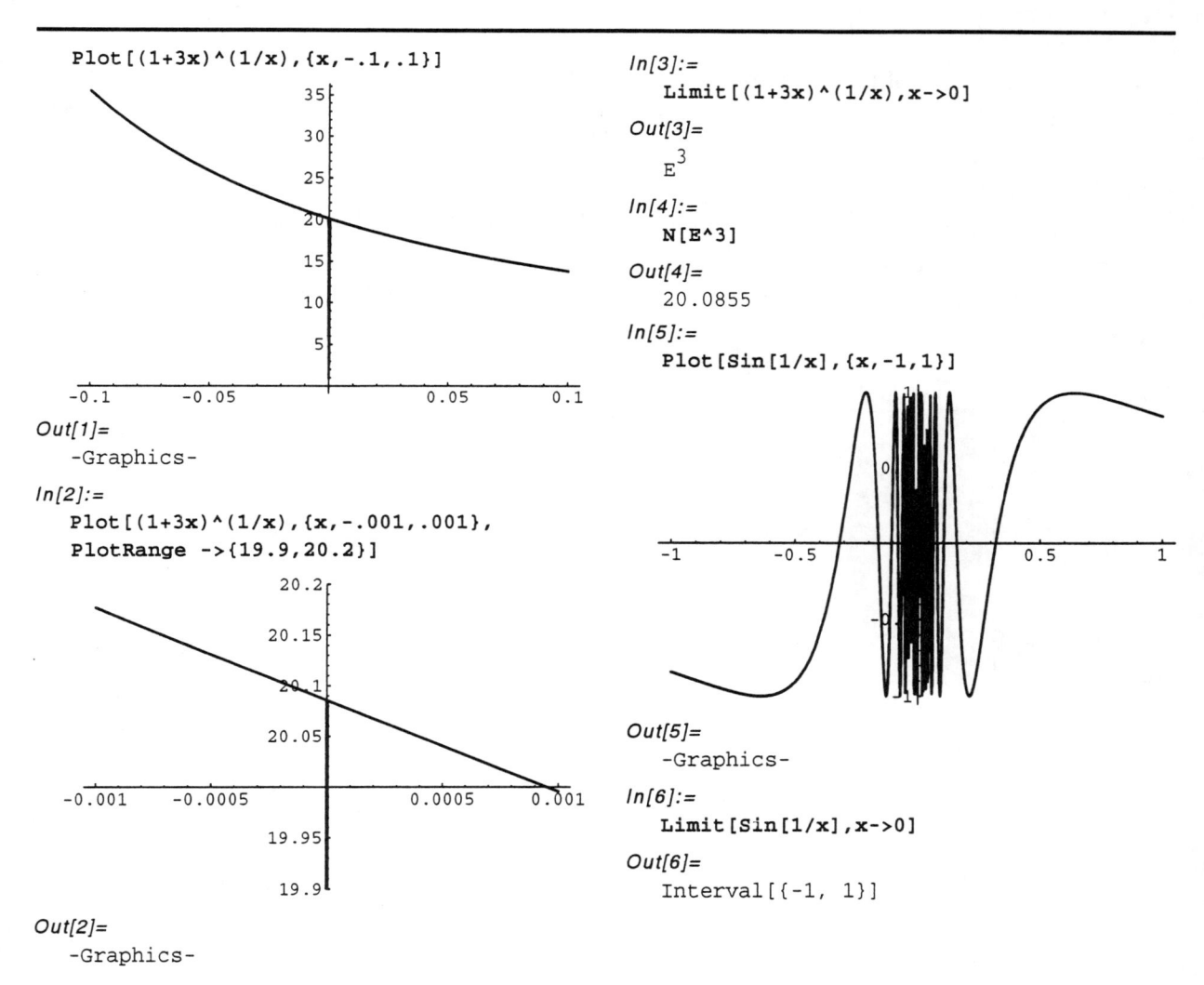

Out[1]=
  -Graphics-

In[2]:=
```
Plot[(1+3x)^(1/x),{x,-.001,.001},
  PlotRange ->{19.9,20.2}]
```

Out[2]=
  -Graphics-

In[3]:=
```
Limit[(1+3x)^(1/x),x->0]
```
Out[3]=
  E^3

In[4]:=
```
N[E^3]
```
Out[4]=
  20.0855

In[5]:=
```
Plot[Sin[1/x],{x,-1,1}]
```

Out[5]=
  -Graphics-

In[6]:=
```
Limit[Sin[1/x],x->0]
```
Out[6]=
  Interval[{-1, 1}]

Figure 2.7: **Limits of** $(1 + 3x)^{\frac{1}{x}}$ **and** $\sin(\frac{1}{x})$

Now we zoom in by changing the domain to $[-0.001, 0.001]$ and the range to $[19.9, 20.2]$ as seen below In[2] of Figure 2.7. The graph crosses the y axis at approximately $(0, 20.086)$. We conclude that when x is near 0, $(1+3x)^{\frac{1}{x}}$ is near 20.086, and this is a reasonable estimate for the limit. (We could improve the accuracy by zooming in.)

Using *MATHEMATICA*'s Limit function, we evaluate **Limit[(1+3x)^(1/x),x->0]** and we obtain e^3, as seen in Out[3] of Figure 2.7. If we numerically approximate this by evaluating **N[E^3]** we get the value 20.0855 and we see that our graphical approximation was very close.

For the second part of the problem, we evaluate **Plot[Sin[1/x],{x,-1,1}]**. The graph is above Out[5] of Figure 2.7. When x is near 0, the graph appears to oscillate wildly and does not get close to any single value. (If you zoom in near the origin, the oscillation will become even more apparent.) We conclude that the limit does not exist.

We can try to evaluate **Limit[Sin[1/x],x->0]**. *MATHEMATICA* responds with Out[6] of Figure 2.7, indicating that no single value is obtained. *Although our assertion that the limit does not exist is correct, this is not a verification; it only means that MATHEMATICA cannot find an answer.*

Laboratory Exercise 2.8

Using Graphs to Estimate Limits (CCH Text 2.7)

Name _____   Due Date _____

Estimate the value of each of the following limits by looking at a graph; then verify your estimate where possible by asking *MATHEMATICA* to calculate the limit.

1. $\lim\limits_{x \to 0} \dfrac{3^x - 1}{x}$

2. $\lim\limits_{x \to \frac{\pi}{2}} \left(x - \dfrac{\pi}{2}\right) \tan x$

3. $\lim\limits_{x \to 0} \dfrac{2^x - 1}{|x|}$

Chapter 3
Key Concept: The Definite Integral

The definite integral is defined as a limit of Riemann sums. Throughout this book all Riemann sums are assumed to use subintervals of equal length.

Solved Problem 3.1: Measuring distance (CCH Text 3.1)

The downward velocity of a parachutist t seconds after jumping from an airplane is given by $v(t) = 20(1 - e^{-1.6t})$ feet per second. Suppose the parachutist jumps from an airplane and lands on the ground 45 seconds later.

(a) Approximate the total distance the parachutist has fallen, using a left-hand Riemann sum with 50 subintervals.

(b) Approximate the total distance the parachutist has fallen, using a right-hand Riemann sum with 50 subintervals.

(c) What is the maximum error in your approximation if you use either of the two estimates above?

(d) What is the maximum error if you use the average of the approximations calculated in Parts (a) and (b) above?

Solution to (a): First we evaluate **Clear[v]** and **v[t_]:= 20(1-E^(-1.6t))**. The left-hand sum for $v(t)$ on $[0, 45]$ with 50 subintervals is $\sum_{i=0}^{49} v(t_i) \Delta t$, where $t_i = \frac{45i}{50}$ and $\Delta t = \frac{45}{50}$. To make the sum, we evaluate **Sum[v[45i/50](45/50),{i,0,49}]**. The Sum function is discussed in Section 9 of Appendix I. The sum appears as Out[3] of Figure 3.1. We see that the value of the left-hand Riemann sum is 876.411 feet.

Solution to (b): The right-hand sum is $\sum_{n=1}^{50} v(t_i) \Delta t$. This time we evaluate **Sum[v[45i/50](45/50),{i,1,50}]** The sum appears in Out[4] of Figure 3.1 and we get 894.411 feet as the value of the right-hand sum.

Solution to (c): In Out[5] of Figure 3.1 we have plotted the graph of $v(t)$. (We used PlotRange->{5,20}.) Since this graph is increasing, the left-hand sum is an underestimate of the total distance traveled, and the right-hand sum is an overestimate. The true distance

```
In[1]:=
    Clear[v]
In[2]:=
    v[t_]:=20(1-E^(-1.6t))
In[3]:=
    Sum[v[45i/50](45/50),{i,0,49}]
Out[3]=
    876.411
In[4]:=
    Sum[v[45i/50](45/50),{i,1,50}]
Out[4]=
    894.411
```

```
In[5]:=
    Plot[v[t],{t,0,45},PlotRange->{5,20}]
```

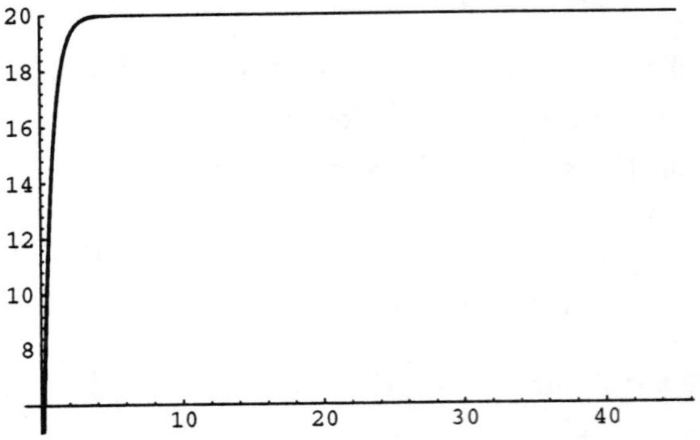

```
Out[5]=
    -Graphics-
```

Figure 3.1: **Riemann sums for the distance fallen by a parachutist**

traveled lies somewhere between the left-hand and right-hand sums calculated in Parts (a) and (b) above. Whether we choose 876.411 or 894.411 as an answer, the error can be no more than their difference, 18.000 feet.

Solution to (d): The average of the left and right sums, 885.411, lies halfway between the two. Since the true distance also lies between the left and right sums, the error can be no more than half their difference, 9.000 feet, when we use the estimate 885.411 feet.

Laboratory Exercise 3.1

An Asteroid (CCH Text 3.1)

Name _____   Due Date _____

An asteroid is falling toward the earth so that its velocity t hours after it was first observed at time $t = 0$ is given by

$$v(t) = \frac{1830000}{(8760 - t)^{\frac{1}{3}}} \text{ kilometers per hour}$$

Determine how far the asteroid travels during the first 6 months after it was first observed. (Take care that your units are correct.)

1. Plot the graph of $v(t)$. (You should adjust the domain to produce a meaningful graph.)

2. Use a left-hand sum with 1000 subintervals to approximate the distance traveled by the asteroid during the first 6 months.

3. Use a right-hand sum with 1000 subintervals to approximate the distance traveled by the asteroid during the first 6 months.

4. Using your work in Parts 2 and 3, make the best estimate you can of the distance the asteroid traveled in the first 6 months. Include an upper bound on the error.

5. When will the asteroid strike the earth? (Hint: Consider the graph in Part 1 and the definition of $v(t)$. They display interesting behavior near a certain point. Discuss what you observe.) Also discuss the validity of the formula for $v(t)$ when the asteroid is very near the earth.

Laboratory Exercise 3.2

A Falling Water Table (CCH Text 3.1)

Name _____   Due Date _____

t days after a heavy rain, the water table at a point near a drainage ditch falls at the rate of

$$v(t) = \frac{10e^{-\frac{1}{t+0.1}}}{(t+0.1)^{\frac{3}{2}}} \text{ inches per day.}$$

1. Use a left-hand Riemann sum with 100 subintervals to estimate how far the water table falls in 3 days.

2. Plot $v(t)$. Can you determine if the number you obtained in Part 1 is an overestimate or an underestimate of the distance the water table has fallen?

3. Use the graph to determine approximately where $v(t)$ is increasing and where it is decreasing. Explain how this information can be used to produce Riemann sums that give an overestimate and an underestimate of the true distance the water table has fallen.

Solved Problem 3.2: Calculating Riemann sums (CCH Text 3.2)

Let $f(x) = x^2$.

(a) Calculate the left-hand Riemann sum for $f(x)$ on $[0, 2]$ using 50 subintervals.

(b) Calculate the right-hand Riemann sum for $f(x)$ on $[0, 2]$ using 50 subintervals.

(c) In order to approximate $\int_0^2 x^2 \, dx$ to one decimal place by a left-hand sum, how many subintervals must be used? What is the left-hand sum for that number of subintervals?

(d) In order to approximate $\int_0^2 x^2 \, dx$ to one decimal place by the average of the left-hand and right-hand sums, how many subintervals must be used? What is this average for that number of subintervals?

(e) Use the MYCALC.m file in Appendix II to plot a graph of $f(x) = x^2$ along with (i) a left-hand sum of 10 subintervals and (ii) a right-hand sum of 10 subintervals.

Solution to (a): The procedure for entering the Riemann sum is the same as that described in Solved Problem 3.1. The sum we want is $\sum_{i=0}^{n-1} f(x_i)\Delta x$, where $f(x) = x^2$, $x_i = \frac{2i}{n}$, $\Delta x = \frac{2}{n}$, and $n = 50$. Evaluate **Clear[f]** and **f[x_]:=x^2**, as in Figure 3.2. Next, we evaluate **Sum[f[2i/50](2/50),{i,0,49}]** and obtain the value 2.5872, as seen in Out[8] of Figure 3.2.

Solution to (b): The right-hand sum is $\sum_{i=1}^{50} f\left(\frac{2i}{50}\right)\frac{2}{50}$. Now we evaluate **Sum[f[2i/50](2/50), {i,1,50}]** and obtain the value 2.7472, as seen in Out[10] of Figure 3.2.

Solution to (c): The key idea is that for a monotone function, the difference between the left-hand and right-hand sums is $|f(b) - f(a)|\Delta x$. Therefore, since the integral is between the two sums, this places a bound on the error between the left-hand sum (or right-hand sum) and the integral. If n is the number of subintervals, then $\Delta x = \frac{2}{n}$, so $|f(2) - f(0)|\Delta x = 4\frac{2}{n} = \frac{8}{n}$. For one decimal place accuracy, we must have $\frac{8}{n} < 0.05$. Thus $n > \frac{8}{0.05} = 160$. If we use 161 subintervals, we can guarantee one-place accuracy.

We need to calculate $\sum_{n=0}^{160} f\left(\frac{2i}{161}\right)\frac{2}{161}$. We evaluate **Sum[f[2i/161](2/161),{i,0,160}]**. The result is displayed in Out[12] of Figure 3.2.

```
In[1]:=
  Clear[f]
In[2]:=
  f[x_]:=x^2
In[7]:=
  Sum[f[2i/50](2/50),{i,0,49}]
Out[7]=
  1617
  ----
  625
In[8]:=
  N[%7]
Out[8]=
  2.5872
In[9]:=
  Sum[f[2i/50](2/50),{i,1,50}]
Out[9]=
  1717
  ----
  625
In[10]:=
  N[%9]
Out[10]=
  2.7472
```

```
In[11]:=
  Sum[f[2i/161](2/161),{i,0,160}]
Out[11]=
  68480
  -----
  25921
In[12]:=
  N[%]
Out[12]=
  2.64187
In[13]:=
  Sum[f[2i/81](2/81),{i,0,80}]
Out[13]=
  51520
  -----
  19683
In[14]:=
  Sum[f[2i/81](2/81),{i,1,81}]
Out[14]=
  53464
  -----
  19683
In[16]:=
  N[(%13+%14)/2]
Out[16]=
  2.66687
```

Figure 3.2: **Riemann sums for** $\int_0^2 x^2\, dx$

Solution to (d): Since the integral is between the two sums and their difference is $|f(b) - f(a)|\Delta x$, then their average is within $\frac{1}{2}|f(b) - f(a)|\Delta x$ of the integral. Therefore, we want n to be large enough to make $\frac{1}{2}|f(b) - f(a)|\Delta x < 0.05$. That is, we need $\frac{1}{2}4\frac{2}{n} < 0.05$, or $n > 80$. We can guarantee one-place accuracy if we use 81 subintervals. The left-hand sum appears in Out[13] of Figure 3.2, and the right-hand sum appears in Out[14]. The average is 2.66687.

Solution to (e): Drawing a Riemann sum with *MATHEMATICA* requires us to type some code that appears in the **Riemann sums** section of Appendix II. We just follow the instructions there.

Assuming the **MYCALC.m** file has been typed and saved in the Packages folder of *MATHEMATICA*, it can be recalled evaluating << **MYCALC.m**. Assuming you defined $f(x) = x^2$ as a function, then evaluate **lhssum[f,{x,0,2,10}]**, as in In[19]. We see the picture above Out[19] of Figure 3.3. This is the left-hand sum.

To get a picture of the right-hand sum, evaluate **rhssum[f,{x,0,2,10}]**, and the right-hand sum appears above Out[20] of Figure 3.3.

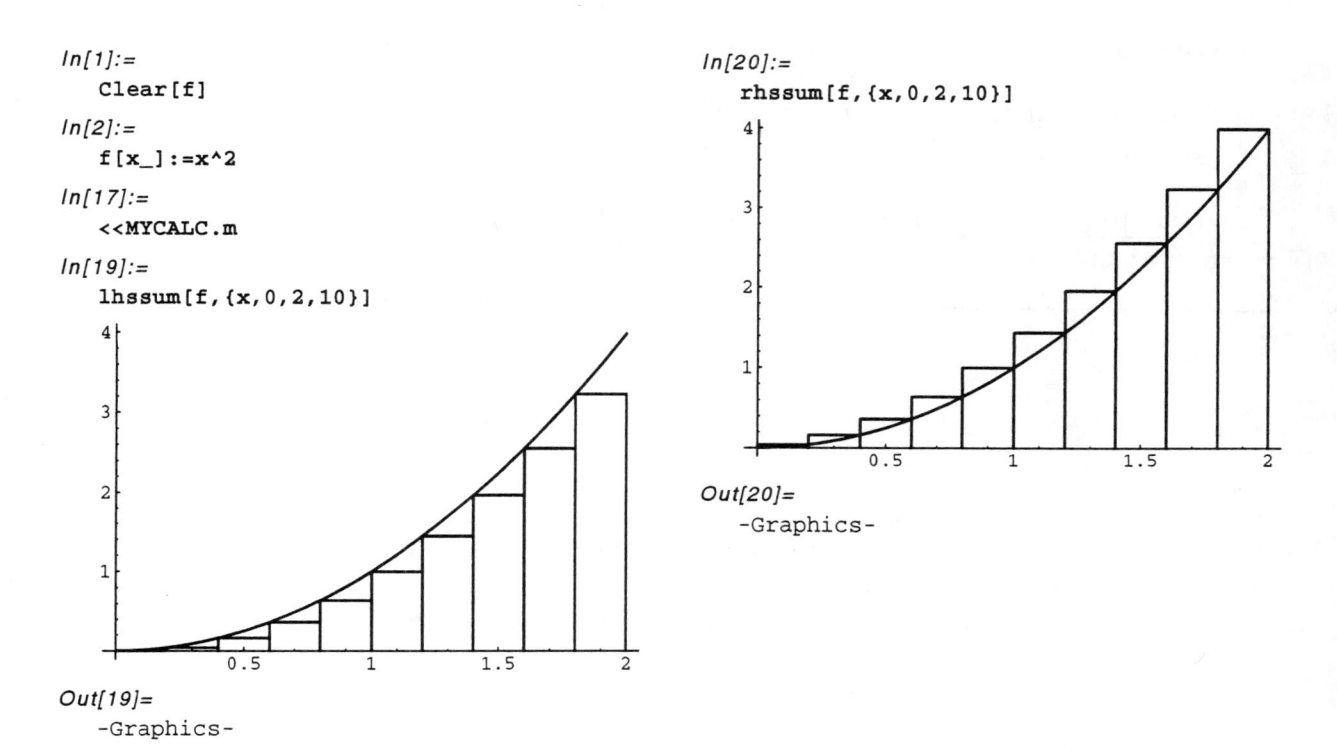

In[1]:=
   Clear[f]

In[2]:=
   f[x_]:=x^2

In[17]:=
   <<MYCALC.m

In[19]:=
   lhssum[f,{x,0,2,10}]

Out[19]=
   -Graphics-

In[20]:=
   rhssum[f,{x,0,2,10}]

Out[20]=
   -Graphics-

Figure 3.3: **Pictures of Riemann sums**

101

Solved Problem 3.3: Limits of Riemann sums (CCH Text 3.2)

Let $f(x) = x^2$.

(a) Calculate the left-hand sum for $f(x)$ on $[1, 2]$ using n subintervals.

(b) Calculate the right-hand sum for $f(x)$ on $[1, 2]$ using n subintervals.

(c) Estimate the limits of the left-hand and the right-hand sums as $n \to \infty$, and explain the meaning of the limits you have calculated.

Solution to (a) and (b): To define the function f, we evaluate **Clear[f]** and **f[x_]:= x^2** (not shown in Figure 3.4).

```
In[20]:=
    lhs[n_]:=Sum[f[1+i/n]*(1/n),{i,0,n-1}]
In[21]:=
    rhs[n_]:=Sum[f[1+i/n]*(1/n),{i,1,n}]
In[24]:=
    lhs[5]
Out[24]=
    51
    ──
    25
In[22]:=
    N[lhs[1000]]
Out[22]=
    2.33183
In[23]:=
    N[rhs[1000]]
Out[23]=
    2.33483
```

Figure 3.4: **Limits of Riemann sums**

The left-hand sum for $f(x)$ on $[1, 2]$ with n subintervals is $\sum_{i=0}^{n-1} f(x_i)\Delta x$, where $x_i = 1 + \dfrac{i}{n}$ and $\Delta x = \dfrac{1}{n}$. Therefore, we evaluate **lhs[n_]:= Sum[f[1+(i/n)]*(1/n),{i,0,n-1}]** and **rhs[n_]:= Sum[f[1+(i/n)]*(1/n),{i,1,n}]** as in In[20] and In[21] of Figure 3.4. If we want the lhs sum with 5 intervals, we evaluate lhs[5] as in Out[24].

<u>Solution to (c)</u>: We now want the limits of these two sums as $n \to \infty$. Evaluate **Limit[lhs[n],n->Infinity]**. Unfortunately, we cannot directly apply *MATHEMATICA*'s Limit function. (This is because as *MATHEMATICA* calculates a limit; it does not use integer values for n and hence the Sum function will not be evaluated.) But we may evaluate the left-hand and right-hand sums for large values of n and get an approximate answer. We evaluate **N[lhs[1000]]** and **N[rhs[1000]]** and get 2.33082 and 2.33483. Since $f(x)$ is increasing, we determine that the exact value of $\int_1^2 x^2$ is between these two values.

Laboratory Exercise 3.3
Calculating Riemann Sums (CCH Text 3.2)

Name _____   Due Date _____

Let $f(x) = \frac{1}{4}(6 + 7x - x^3)$.

1. Calculate the left-hand Riemann sum for $f(x)$ on $[-1, 1]$ using 30 subintervals.

2. Calculate the right-hand Riemann sum for $f(x)$ on $[-1, 1]$ using 30 subintervals.

3. In order to approximate $\int_{-1}^{1} f(x)\, dx$ to one decimal place by a right-hand sum, how many subintervals must be used? What is the right-hand sum for that number of subintervals?

4. In order to approximate $\int_{-1}^{1} f(x)\, dx$ to one decimal place by the average of the left-hand and right-hand sums, how many subintervals must be used? What is this average for that number of subintervals?

5. Use the **MYCALC.m** file in Appendix II to plot a graph of $f(x)$ on $[-1, 3]$ with a left-hand sum of 10 subintervals. In Solved Problem 3.2 the left-hand sum stayed underneath the graph. Did that happen here? Explain.

6. Use the **MYCALC.m** file in Appendix II to plot a graph of $f(x)$ on $[-1, 3]$ with a right-hand sum of 10 subintervals. In Solved Problem 3.2 the right-hand sum stayed above the graph. Did that happen here? Explain.

Laboratory Exercise 3.4

Limits of Riemann Sums (CCH Text 3.2)

Name _____   Due Date _____

Let $f(x) = \frac{1}{4}(6 + 7x - x^3)$.

1. Calculate the left-hand sum for $f(x)$ on [2, 3] using n subintervals.

2. Calculate the right-hand sum for $f(x)$ on [2, 3] using n subintervals.

3. Approximate the limits of the left-hand and the right-hand sums as $n \to \infty$, and explain the meaning of the limits you have calculated.

Laboratory Exercise 3.5

Estimating Integrals with Riemann Sums (CCH Text 3.2)

Name _____   Due Date _____

In each of the following, (a) find the number of subintervals necessary for the average of the left and right sums to approximate the integral to one decimal place, and (b) find the average of the left and right sums for that number of subintervals.

1. $\int_1^2 \sin(\ln t)\, dt$

2. $\int_{\pi/2}^{\pi} \sqrt{\sin t}\, dt$

Laboratory Exercise 3.6

Riemann Sums and the Fundamental Theorem (CCH Text 3.4)

Name _____   Due Date _____

1. Calculate the left-hand sum for $f(t) = t^2$ on $[0, x]$ using n subintervals.

2. Approximate the limit of the left-hand sum above as $n \to \infty$.

3. Calculate $\int_0^x t^2 \, dt$ using the Fundamental Theorem of Calculus.

4. What do you observe about your answers in Parts 2 and 3?

5. Repeat Parts 1 through 4 above for $f(t) = e^t$.

Calculating Definite Integrals with *MATHEMATICA* (CCH Text 3.4)

MATHEMATICA can calculate many definite integrals exactly using the Fundamental Theorem of Calculus. We will illustrate the procedure with $\int_0^\pi \sin x \, dx$. We will use the *MATHEMATICA* function Integrate. Evaluating **Integrate[Sin[x],{x,0,Pi}]** causes *MATHEMATICA* to try to integrate the function $\sin x$ with respect to x from 0 to π, as seen in Out[1] in Figure 3.5.

```
In[1]:=
    Integrate[Sin[x],{x,0,Pi}]
Out[1]=
    2
In[2]:=
    Integrate[Sin[x^2],{x,0,Pi}]
Out[2]=
    Sqrt[Pi/2] FresnelS[Sqrt[2 Pi]]
In[3]:=
    NIntegrate[Sin[x^2],{x,0,Pi}]
Out[3]=
    0.772652
```

Figure 3.5: *MATHEMATICA*'s **calculation of** $\int_0^\pi \sin x \, dx$

How could *MATHEMATICA* do that? *MATHEMATICA* knows that $\dfrac{d}{dx} - \cos x = \sin x$. Thus, applying the Fundamental Theorem of Calculus gives

$$\int_0^\pi \sin x \, dx = (-\cos \pi) - (-\cos 0) = 2$$

The actual rules *MATHEMATICA* uses are more complicated for definite integrals and involve infinite series.

MATHEMATICA can calculate many definite integrals exactly, but sometimes it fails. For example, let's look at $\int_0^\pi \sin(x^2) \, dx$. Evaluate **Integrate[Sin[x^2],{x,0,Pi}]**. See Out[2] of Figure 3.5. This indicates that *MATHEMATICA* is unable to calculate the integral exactly; it does not know a function whose derivative is $\sin(x^2)$, and so cannot apply the Fundamental

Theorem of Calculus. *MATHEMATICA* should be forgiven for this lapse, because there is no "simple" function whose derivative is $\sin(x^2)$.

Even though *MATHEMATICA* cannot calculate $\int_0^\pi \sin(x^2)\,dx$ exactly, it knows how to calculate an approximation with the function NIntegrate. We evaluate **NIntegrate[Sin[x^2],{x,0,Pi}]**. The result is displayed in Out[3] of Figure 3.5.

Laboratory Exercise 3.7

Calculating Areas (CCH Text 3.3)

Name _____   Due Date _____

Produce pictures of the following regions and calculate their areas. Express your answers as integrals and ask *MATHEMATICA* to evaluate them. Where exact answers cannot be found, provide approximations.

1. The region between $\sin x$ and the $x$axis from $x = 0$ to $x = 2\pi$. (Be careful. The correct answer is *not* $\int_0^{2\pi} \sin x \, dx$. Remember that the integral gives area over intervals where the function is positive, but gives the negative of the area over intervals where the function is negative.)

2. The region inside the unit circle above the graph of $y = x^2$.

3. The region inside the unit circle above the graph of $y = x^3$. (Hint: If you examine the picture carefully, you can find the exact answer without calculus or a computer.)

Laboratory Exercise 3.8

The Average Value of a Function and the Fundamental Theorem
(CCH Text 3.4)

Name _____   Due Date _____

1. Find the average value of x^4 on $[a, b]$. <u>Hint</u>: *MATHEMATICA* can symbolically integrate functions. Try evaluating **Integrate[x^4,x,a,b]**.

2. Find the limit of your answer in Part 1 as $b \to a$.

3. Use *MATHEMATICA* to try to find the average value of $\sin(x^2)$ on the interval $[a, b]$. What does *MATHEMATICA*'s response tell you about its ability to find this integral symbolically? For specific a and b we can approximate the average. Do this approximation for $a = 1$ and $b = 1.5$.

4. Using *MATHEMATICA*'s Limit function, calculate the limit of the integral in Part 3 as $b \to a$.

5. In Part 2, it was not hard to see how *MATHEMATICA* could find the limit, because it could find an explicit formula for the integral. However, even though it could not evaluate the integral in Part 3, it was still able to find the limit of the average values in Part 4! Ask *MATHEMATICA* to find $\lim_{b \to a} \dfrac{\int_a^b f(x)\, dx}{b - a}$. What did you get?

6. By using graphs and the concept of "average value of a function," explain how *MATHEMATICA* is able to get the answers to Part 5.

Chapter 4
Short Cuts to Differentiation

MATHEMATICA can find the derivative of virtually any combination of the usual elementary functions. Its answers may occasionally look different from the ones you get by hand, but that doesn't mean you're wrong. Just try simplifying a bit more.

Solved Problem 4.1: Difference quotients and derivatives (CCH Text 4.2)

(a) Use *MATHEMATICA* to verify that $\frac{d}{dx}x^{10} = 10x^9$.

(b) Use *MATHEMATICA* to simplify the difference quotient $\frac{(x+h)^{10} - x^{10}}{h}$ and explain how the simplified form can be used to verify that $\frac{d}{dx}x^{10} = 10x^9$.

Solution to (a): The point of this part of the exercise is merely to show the steps involved in using *MATHEMATICA* to calculate derivatives. Enter and evaluate the two input lines **Clear[f]** and **f[x_] := x^10** as seen in In[1] and In[2] of Figure 4.1.

```
In[1]:=
    Clear[f]
In[2]:=
    f[x_]:=x^10
In[3]:=
    D[f[x],x]
Out[3]=
    10 x^9
```

```
In[4]:=
    Clear[g]
In[5]:=
    g[h_]:= ((x+h)^10-x^10)/h
In[6]:=
    Limit[g[h],h->0]
Out[6]=
    10 x^9
```

Figure 4.1: **The derivative of x^{10}**

Now use the differentiate function **D[ ]** to find the derivative. Evaluating **D[f[x],x]** causes *MATHEMATICA* to symbolically differentiate the function $f(x)$ with respect to x. That is, *MATHEMATICA* will output $\frac{df}{dx}$. See In[3] and Out[3] of Figure 4.1.

Solution to (b): We evaluate **Clear[g]** and **g[h_] := ((x+h)^10-x^10)/h**. Notice the use of the parentheses to assure that we divide the difference by h and not just x^10. The result is displayed in In[5] of Figure 4.1. The derivative of x^{10} is the limit as h goes to zero of this expression, and each term except the first of the simplified expression is multiplied by some power of h. Evaluate **Limit[g[h],h -> 0]**. In Figure 4.1 the limit appears in Out[6]. We note that Out[3] and Out[6] are identical.

Laboratory Exercise 4.1

Difference Quotients and the Derivative (CCH Text 4.2)

Name _____   Due Date _____

For each of the following functions, (a) use *MATHEMATICA* to calculate the derivative directly, and (b) ask *MATHEMATICA* to simplify the appropriate difference quotient, and explain how it can be used to verify *MATHEMATICA*'s answer from Part (a).

1. $f(x) = \dfrac{1}{x^5}$

2. $f(x) = \dfrac{x}{x+1}$

3. $f(x) = \dfrac{x^2}{x+5}$

4. $f(x) = \sqrt{x}$

Solved Problem 4.2: The product rule (CCH Text 4.4)

Verify the product rule and the Fundamental Theorem of Calculus with *MATHEMATICA*.

Solution: The point of this exercise is not really to prove that *MATHEMATICA* knows the product rule and the Fundamental Theorem of Calculus, but to show you how to use *MATHEMATICA* to work with derivatives of arbitrary functions.

Evaluate **Clear[f,g]** and **D[f[x]*g[x],x]** as seen in Out[2] of Figure 4.2. This is the product rule.

```
In[1]:=
   Clear[f,g]
In[2]:=
   D[f[x]*g[x],x]
Out[2]=
   g[x] f'[x] + f[x] g'[x]
In[4]:=
   Integrate[D[f[x],x],x]
Out[4]=
   f[x]
```

Figure 4.2: **The product rule and the Fundamental Theorem of Calculus**

To illustrate the Fundamental Theorem of Calculus, we will integrate the derivative of $f(x)$. Evaluate **Integrate[D[f[x],x], x]**. We see from Out[4] that *MATHEMATICA* does indeed know the Fundamental Theorem of Calculus.

Laboratory Exercise 4.2

The Quotient Rule and Chain Rule (CCH Text 4.5)

Name _____   Due Date _____

1. Ask *MATHEMATICA* to verify the quotient rule by differentiating $f(x)/g(x)$. Is *MATHEMATICA*'s answer the same as that in your textbook? If not, show that the two are the same.

2. Find the derivative of $f^2(x)$ by hand, and then ask *MATHEMATICA* to find its derivative. (Remember to enter $f^2(x)$ as f[x]^2.) Are your answers the same?

3. Find the derivative of $f(x^2)$ by hand, and then ask *MATHEMATICA* to find its derivative. Are your answers the same?

4. Find the derivative of $\sin(f(x))$ by hand, and then ask *MATHEMATICA* to find its derivative. Are your answers the same?

5. Find the derivative of $f(\sin(x))$ by hand, and then ask *MATHEMATICA* to find its derivative. Are your answers the same?

6. Based on your observations in the last four parts, does *MATHEMATICA* seem to know about the chain rule?

7. Ask *MATHEMATICA* to calculate the derivative of $\int_a^x f(t)\, dt$. Prove that *MATHEMATICA*'s answer is correct. (<u>Hint</u>: Let $f(t) = g'(t)$.)

Solved Problem 4.3: Implicit differentiation (CCH Text 4.8)

Suppose $\sin(x+y) + \sqrt{x^2 + y^2} = xy + 4$.

(a) Find $\dfrac{dy}{dx}$ by implicit differentiation.

(b) Find $\dfrac{dy}{dx}$ when $x = \sqrt{2}$ and $y = -\sqrt{2}$.

(c) Find the equation of the tangent line to the graph of the given equation at $(\sqrt{2}, -\sqrt{2})$. Use it to approximate the value of y when $x = 1.5$.

Solution to (a): The first step is to tell *MATHEMATICA* that y is a function of x by evaluating **Clear[f]** and **y=f[x]**. (If we do not do this, *MATHEMATICA* will treat y as a constant when we differentiate rather than as a function of x.) To differentiate, we evaluate **D[Sin[x+y]+Sqrt[x^2+y^2]-x*y -4,x]**, as seen in Outputs 1-3 of Figure 4.3. Alternatively, you could use **y[x]** instead of y and *MATHEMATICA* would know that y is a function of x.

Notice that *MATHEMATICA* has replaced y by $f(x)$. We wish to find $\dfrac{d}{dx} f(x)$. So, to find y', evaluate **Solve[% == 0, f'[x]]**. We obtain the result in Out[4] of Figure 4.3. To replace $f(x)$ with y again, we need to evaluate **Clear[y]**. Now we replace $f(x)$ with y by evaluating **%4/.(f[x]->y)**. (Remember %n is a fast way of using Out[n].) The solution appears as Out[6] in Figure 4.4.

Solution to (b): (Before we start, we observe that the given values $x = \sqrt{2}$ and $y = -\sqrt{2}$ really do satisfy the given equation.) We enter the implicit equation as follows. Evaluate **Clear[g,y]** and **g[x_,y_]:= Sin[x+y]+Sqrt[x^2+y^2]-x*y -4**. By evaluating $g(\sqrt{2}, -\sqrt{2})$ we get 0, which is what we want. See Output[4] in Figure 4.4. Now we can differentiate again and evaluate the derivative at the desired point. Again we evaluate **D[g[x,f[x]],x]/.f[x]->y**. Next we replace x by $\sqrt{2}$ and y by $-\sqrt{2}$ by evaluating **%/.{x->Sqrt[2],y->-Sqrt[2]}**. See Out[6] in Figure 4.4. Now we can find $f'(\sqrt{2})$ by evaluating **Solve[%==0,f'[Sqrt[2]]]** as in Out[7]. We see that the derivative is $-\dfrac{2+3\sqrt{2}}{2-3\sqrt{2}}$.

Solution to (c): The tangent line passes through the point $(\sqrt{2}, -\sqrt{2})$ and has slope $-\dfrac{2+3\sqrt{2}}{2-3\sqrt{2}}$. Thus its equation is $y = -\dfrac{2+3\sqrt{2}}{2-3\sqrt{2}}(x - \sqrt{2}) - \sqrt{2}$. The value $x = 1.5$ is close to $\sqrt{2}$, and so the tangent line should yield a good approximation to the true function value. The result of this is displayed in Out[8] of Figure 4.4.

Remark: One may check the accuracy by replacing x by 1.5 in the original equation and solving for y. Try it.

In[1]:=
   Clear[f,y]

In[2]:=
   y=f[x]

Out[2]=
   f[x]

In[3]:=
   D[Sin[x+y]+Sqrt[x^2+y^2]-x*y-4,x]

Out[3]=
$$-f[x] - x\,f'[x] + \cos[x + f[x]]\,(1 + f'[x]) + \frac{2x + 2f[x]\,f'[x]}{2\,\text{Sqrt}[x^2 + f[x]^2]}$$

In[4]:=
   Solve[%==0,f'[x]]

Out[4]=
$$\{\{f'[x] \to -((x + \cos[x + f[x]]\,\text{Sqrt}[x^2 + f[x]^2] - f[x]\,\text{Sqrt}[x^2 + f[x]^2]) / (f[x] - x\,\text{Sqrt}[x^2 + f[x]^2] + \cos[x + f[x]]\,\text{Sqrt}[x^2 + f[x]^2]))\}\}$$

In[5]:=
   Clear[y]

In[6]:=
   %4/.(f[x]->y)

Out[6]=
$$\{\{f'[x] \to -(\frac{x - y\,\text{Sqrt}[x^2 + y^2] + \text{Sqrt}[x^2 + y^2]\,\cos[x + y]}{y - x\,\text{Sqrt}[x^2 + y^2] + \text{Sqrt}[x^2 + y^2]\,\cos[x + y]})\}\}$$

Figure 4.3: **Implicit differentiation**

In[1]:=
  Clear[g,y]

In[2]:=
  g[x_,y_]:=Sin[x+y]+Sqrt[x^2+y^2]-x*y-4

In[4]:=
  g[Sqrt[2],-Sqrt[2]]

Out[4]=
  0

In[5]:=
  D[g[x,f[x]],x]/.f[x]->y

Out[5]=
$$-y - x\, f'[x] + \cos[x + y]\,(1 + f'[x]) + \frac{2x + 2y\, f'[x]}{2\, \text{Sqrt}[x^2 + y^2]}$$

In[6]:=
  %/.{x->Sqrt[2],y->-Sqrt[2]}

Out[6]=
$$1 + \text{Sqrt}[2] + f'[\text{Sqrt}[2]] - \text{Sqrt}[2]\, f'[\text{Sqrt}[2]] + \frac{2\,\text{Sqrt}[2] - 2\,\text{Sqrt}[2]\, f'[\text{Sqrt}[2]]}{4}$$

In[7]:=
  Solve[%==0,f'[Sqrt[2]]]

Out[7]=
$$\{\{f'[\text{Sqrt}[2]] \to -(\frac{2 + 3\,\text{Sqrt}[2]}{2 - 3\,\text{Sqrt}[2]})\}\}$$

In[8]:=
  N[-(2+3Sqrt[2])/(2-3Sqrt[2])(1.5 - Sqrt[2])-Sqrt[2]]

Out[8]=
  -1.17542

Figure 4.4: **Implicit differentiation continued**

Laboratory Exercise 4.3

Implicit Differentiation (CCH Text 4.8)

Name _____   Due Date _____

Assume y is a function of x such that $ye^{xy^2} + ye^x + xe^y = 2$.

1. Find $\dfrac{dy}{dx}$ by implicit differentiation.

2. Verify that $x = 0$ and $y = 1$ satisfy the given equation.

3. Find $\dfrac{dy}{dx}$ when $x = 0$ and $y = 1$.

4. Find the equation of the tangent line at $(0,1)$, and use it to approximate the value of $y(0.15)$.

5. Check the accuracy of your approximation in Part 4 by substituting $x = 0.15$ into the original equation, and solving for y.

Chapter 5
Using the Derivative

Since the derivative measures the rate of change of a function, it can help solve many different problems, especially those that involve determining where a function reaches a maximum or minimum.

A word of advice concerning word problems: When the problem involves physical units, be sure your final answer does too. Also, always ask yourself whether your answer makes sense. If you are supposed to find the length of a pencil, and you get an answer such as 200 miles, -0.22 meters, or $3 + i\sqrt{5}$ feet, you should suspect something is wrong.

Solved Problem 5.1: Maxima and minima (CCH Text 5.1)

Let $f(x) = e^{2x} - 3e^x$.

(a) Use the graph of f to estimate all local maxima and minima of f.

(b) Plot the graph of f' and explain how this graph supports your conclusions in Part (a).

(c) Use the derivative of f to find the *exact* values of the maxima and minima.

Solution to (a): We evaluate **Clear[f]** and **f[x_]:= E^(2x)-3E^x**. Evaluate **Plot[f[x], { x, -2, 2 }]** to obtain the graph of f as seen above Out[3] of Figure 5.1. We see that the graph has a single local minimum, which *appears* to also be a global minimum, and there *seems* to be no local maximum. Notice that we are careful not to make the last two assertions definitively, because we aren't yet sure whether something may be happening off the computer screen to the contrary. Part (c) will resolve this.

We use the mouse to enlarge the graphics window and to find the approximate coordinates of the bottom point. The minimum point is approximately $(0.403, -2.266)$.

Solution to (b): We return to our calculations and differentiate $f(x)$. The derivative appears in Out[4] of Figure 5.1. To add the graph of the derivative to the graph of $f(x)$, we evaluate **Clear[df]**, **df[_]:= -3E^x + 2E^(2x)** and **Plot[{f[x],df[x]},{x,0,1},PlotStyle -> {Thickness[0.001],Thickness[0.009]}]**. (Alternatively, we could have gotten $f'(x)$ by **df[x_]:= D[f[t],t]/.t->x**.) We observe that the graph of the derivative crosses the x axis, changing from negative to positive at about $x = 0.4$. This shows that f decreases up to this point, reaches a local minimum, and then starts to increase. (Accuracy could be further improved by zooming in.)

```
In[1]:=
   Clear[f]
In[2]:=
   f[x_]:=E^(2x)-3E^x
In[3]:=
   Plot[f[x],{x,-2,2}]
```

Out[3]=
  -Graphics-

```
In[4]:=
   D[f[x],x]
```

Out[4]=
  $-3 E^x + 2 E^{2x}$

```
In[5]:=
   Clear[df]
In[6]:=
   df[x_]:=-3E^x+2E^(2x)
```

```
In[7]:=
   Plot[{f[x],df[x]},{x,0,1},PlotStyle->
     {Thickness[0.001],Thickness[0.009]}]
```

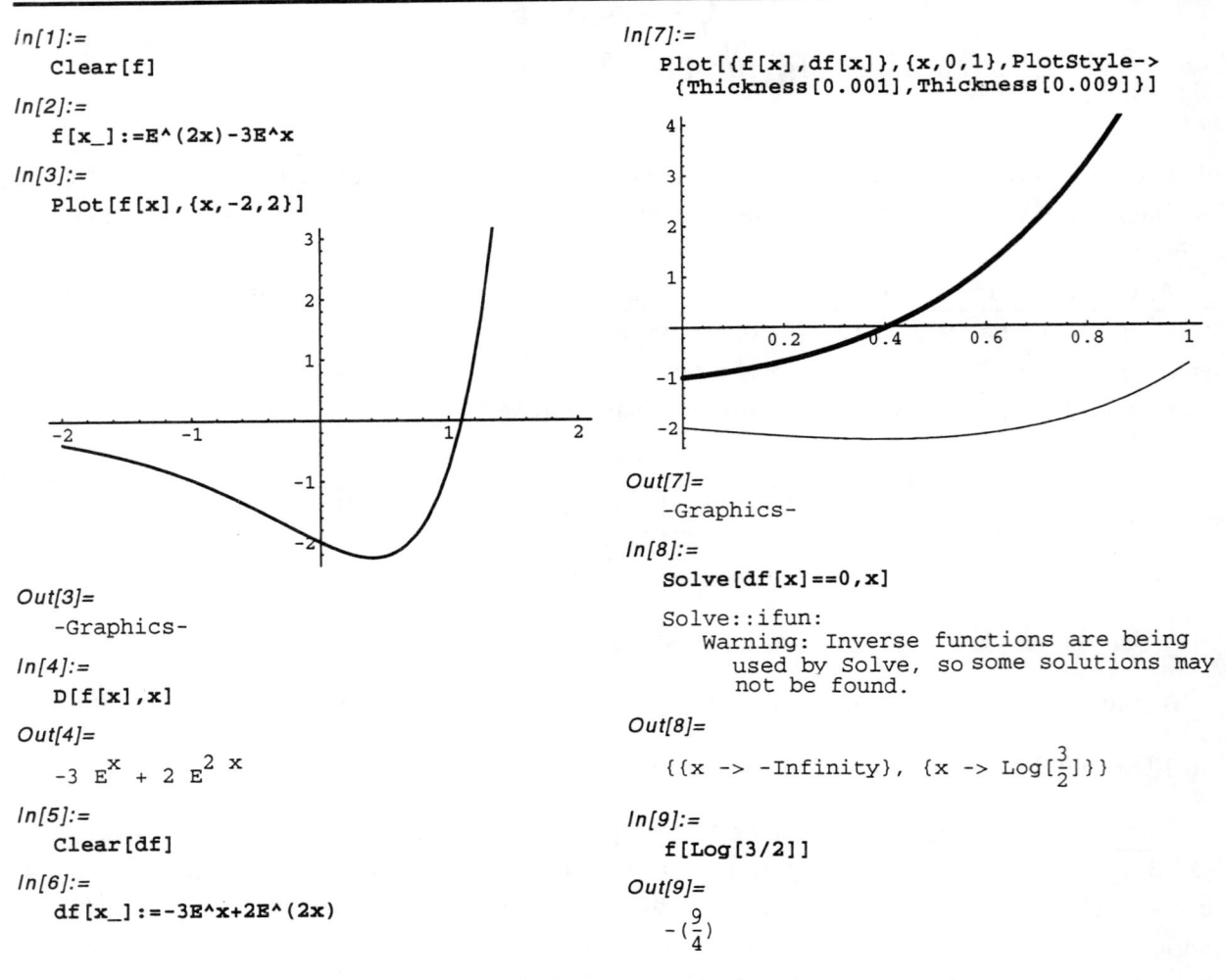

Out[7]=
  -Graphics-

```
In[8]:=
   Solve[df[x]==0,x]
```

Solve::ifun:
   Warning: Inverse functions are being
     used by Solve, so some solutions may
     not be found.

Out[8]=
  $\{\{x \to -\text{Infinity}\}, \{x \to \text{Log}[\frac{3}{2}]\}\}$

```
In[9]:=
   f[Log[3/2]]
```

Out[9]=
  $-(\frac{9}{4})$

Figure 5.1: **Extrema of $e^{2x} - 3e^x$**

Solution to (c): To get the exact value of the minimum, we must find the critical values of f, that is, the zeros of its first derivative. We evaluate **FindRoot[df[x]==0,{x,0.4}]** to get an approximate value or **Solve[df[x]==0,x]** as in of Figure 5.1 to get the single critical value, $\ln(\frac{3}{2})$. Because this is the *only* critical point, we can now be sure that the graph indeed has a global minimum at this point and that there is no local maximum.

To find the y value, evaluate **f[Log[3/2]]**, which yields $-\frac{9}{4}$ as in Out[9] of Figure 5.1. (You should be able to perform this simplification by hand. Try it.) By using the numerical approximation function N[ ] we could approximate the x and y coordinates of the minimum, so they can be compared with our estimates in Part (a).

> **Solved Problem 5.2: Critical points and extrema (CCH Text 5.1)**

Let $f(x) = |x^4 + x - 1| - x$.

(a) Plot both f and f' on the same screen.

(b) Locate each critical value of f and estimate each local maximum and minimum correct to six decimal places.

(c) Find the global maximum and minimum values of f.

<u>Solution to (a)</u>: First we evaluate **Clear[f]** and **f[x_]:= If[x^4 +x-1>0,x^4-1,-x^4-2x+1]**. This uses *MATHEMATICA*'s If construction. The syntax is **If[***condition***,t,f]**. If *condition* is true, then t is evaluated, and if *condition* is false, then f is evaluated. *MATHEMATICA* has an absolute value function, Abs[ ], but this does not work well with differentiation. We want the derivative as a function. We evaluate **Clear[df]** and **D[f[x],x]**. We create a function by copying this output and evaluating **df[x_]:= If[x^4+x-1>=0,4x^3,-2-4x^3]**, as in In[4] of Figure 5.2. We plot these functions by evaluating **Plot[{f[x],df[x]},{x,-2,3},PlotStyle-> {Thickness[0.001],Thickness[0.008]}]**, as in Out[5] of Figure 5.2.

Note that the graph of f shows two local minima and one local maximum. Notice that the jumps in the derivative are shown as dark vertical lines. This is caused by the way the program makes graphs by connecting dots with line segments.

<u>Solution to (b)</u>: A *critical value* is a value of x where the derivative of f is zero or undefined. From the graph of the derivative we clearly see three critical values: one negative value where the graph crosses the x axis and two discontinuities where it is undefined. The discontinuities are where $x^4 + x - 1$ changes sign. To find these values, we evaluate **FindRoot[x^4+x-1,{x,-1.2}]** and **FindRoot[x^4+x-1,{x,.7}]**. By evaluating $f(x)$ at these points, we get that the local minima are $(-1.22072, 1.22074)$ and $(0.724492, -0.724492)$. See outputs 6 and 7. To find the local maximum, we evaluate **FindRoot[df[x]==0,{x,-.7}]**. After evaluating $f(x)$ at this point we see that the local maximum is $(-0.793701, 2.19055)$, as in outputs 8 and 9 in Figure 5.2.

By zooming in on the graph we can also get estimates, but the method above is more accurate.

(c) There is no global maximum, but $y = -0.724492$ is a global minimum occurring at $x = 0.724492$.

```
In[1]:=
    Clear[f]
In[2]:=
    f[x_]:=If[x^4+x-1>=0,x^4-1,-x^4-2x+1]
In[3]:=
    Clear[df]
In[4]:=
    df[x_]:=If[x^4+x-1>=0,4x^3,-4x^3-2]
In[5]:=
    Plot[{f[x],df[x]},{x,-2,3},PlotStyle->
     {Thickness[0.001],Thickness[0.008]}]
```

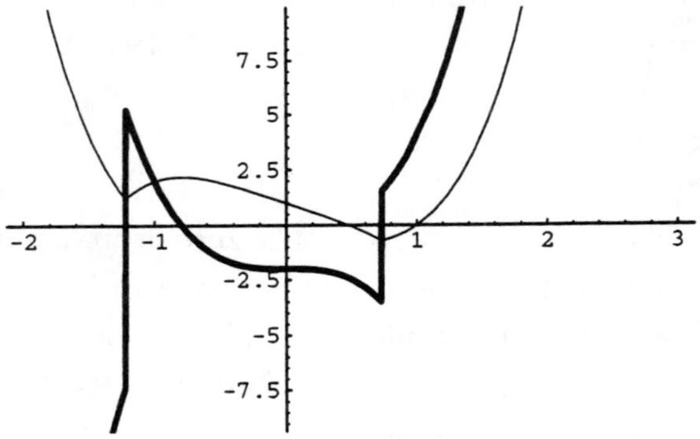

Out[5]=
    -Graphics-

```
In[6]:=
    FindRoot[x^4+x-1==0,{x,-1.2}]
Out[6]=
    {x -> -1.22074}
In[7]:=
    FindRoot[x^4+x-1==0,{x,.7}]
Out[7]=
    {x -> 0.724492}
In[8]:=
    FindRoot[df[x]==0,{x,-.7}]
Out[8]=
    {x -> -0.793701}
In[9]:=
    f[-0.793701]
Out[9]=
    2.19055
```

Figure 5.2: $|x^4 + x - 1| - x$ and its derivative

Laboratory Exercise 5.1

Local Extrema (CCH Text 5.1)

Name _____   Due Date _____

Find all the critical values of each function f below to six decimal place accuracy, and find all the local (or relative) extrema, as well as the absolute extrema of f. Explain how you arrive at each of your answers. Plot the graph of f and its derivative so that all local extrema are shown.

1. $f(x) = x^2 + \sin x$

2. $f(x) = \dfrac{x^2 + 1}{x + |x^3 + x - 1|}$

Solved Problem 5.3: Inflection points (CCH Text 5.2)

Plot the graphs of $f(x) = x^6 - x^4$ and its second derivative on the same screen. Find the inflection points of f.

Solution: We evaluate **Clear[f,df,ddf]** and **f= x^6-x^4**. Note that we used f instead of $f(x)$, since we will only be graphing f. Next evaluate **df = D[f,x]** and then **ddf = D[f,{x,2}]**. The latter is using the syntax D[f,{x,n}], where this evaluates to the n th derivative of the function f with respect to x. (We could have just as easily used **dff = D[df,x]**. See outputs 2-4 of Figure 5.3. Next evaluate **Plot[{ f, df, dff }, {x ,- 2, 2}, PlotStyle -> {Thickness[0.001]**,

```
In[13]:=
    Clear[f,df,ddf]

In[14]:=
    f=x^6-x^4
Out[14]=
    -x^4 + x^6

In[15]:=
    df=D[f,x]
Out[15]=
    -4 x^3 + 6 x^5

In[16]:=
    ddf=D[f,{x,2}]
Out[16]=
    -12 x^2 + 30 x^4
```

```
In[17]:=
    Plot[{f,df,ddf},{x,-2,2},PlotStyle->
        {Thickness[0.001],Thickness[0.005],Thickness[0.009]}]
```

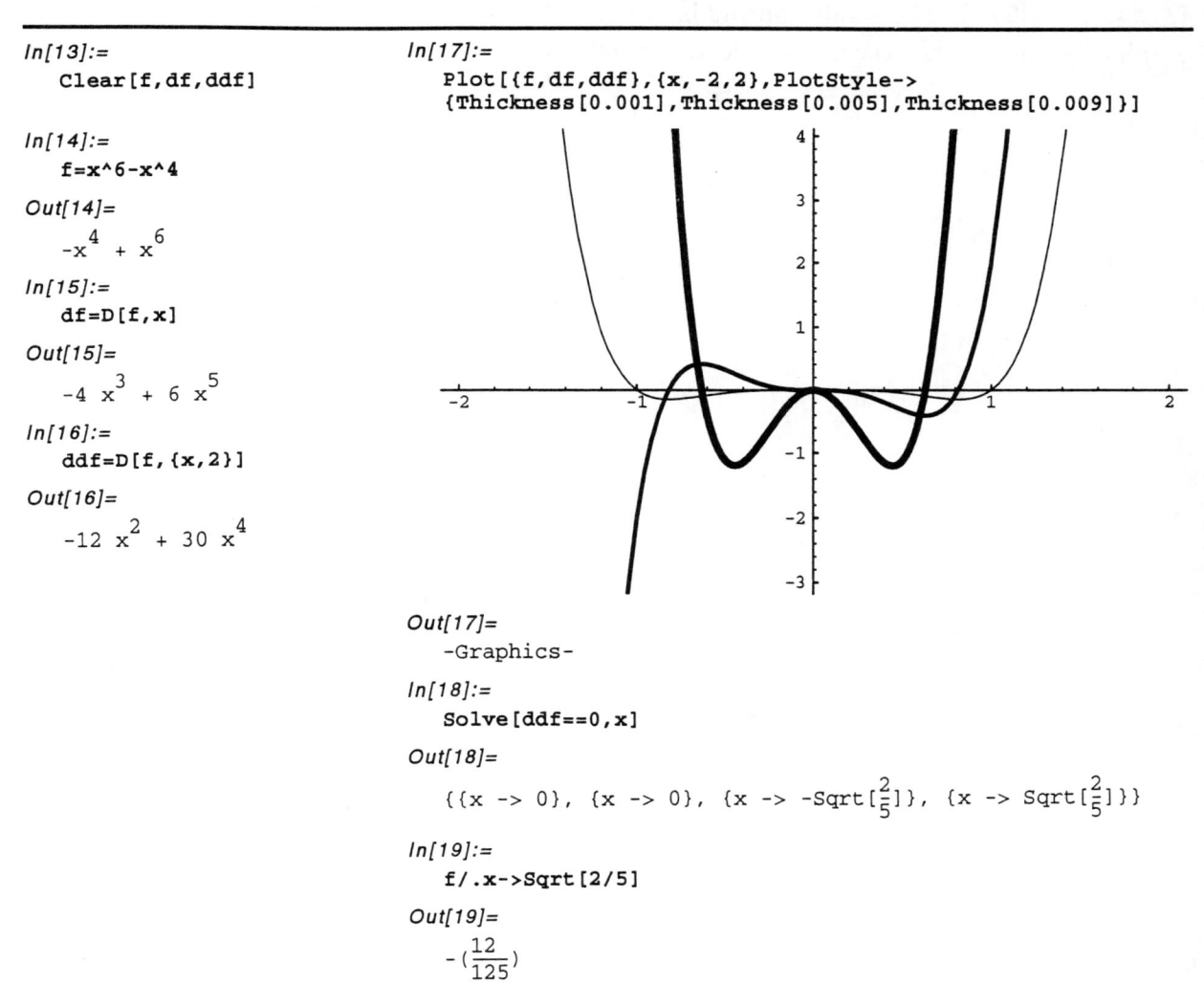

```
Out[17]=
    -Graphics-

In[18]:=
    Solve[ddf==0,x]
Out[18]=
    {{x -> 0}, {x -> 0}, {x -> -Sqrt[2/5]}, {x -> Sqrt[2/5]}}

In[19]:=
    f/.x->Sqrt[2/5]
Out[19]=
    -(12/125)
```

Figure 5.3: **Inflection points for** $x^6 - x^4$

139

Thickness[0.005],Thickness[0.008]}]. Enlarge the graph using the mouse at the edge of the graphics window. See Out[17].

Points of inflection may occur where the second derivative is zero. Therefore, evaluate **Solve[dff == 0,x]** and *MATHEMATICA* presents us with the four roots (with $x = 0$ occurring twice, in Out[18] of Figure 5.3.

The graph of the second derivative crosses the x axis, indicating a change in concavity in the graph of f at both $x = -\frac{\sqrt{10}}{5}$ and $x = \frac{\sqrt{10}}{5}$. Thus, these are the x coordinates of inflection points. On the other hand, since the graph of the second derivative does not cross the x axis at $x = 0$, no change in concavity occurs there. We conclude that $(0, 0)$ is *not* an inflection point.

To find the y coordinates of the inflection points, we evaluate f at one of these points by **f/.x -> Sqrt[2/5]**. The result appears in Out[19]. We obtain $y = -\frac{12}{125}$. The symmetry of the graph of f tells us that the other inflection point has this same y coordinate.

Laboratory Exercise 5.2

Inflection Points (CCH Text 5.2)

Name _____   Due Date _____

For each of the following functions, plot the graphs of f and f'', and find all the inflection points of f.

1. $f(x) = \arctan(1 + x^2)$

2. $f(x) = x^2 e^{-x}$

3. $f(x) = x^4 \ln(1 + e^x)$

Laboratory Exercise 5.3

The George Deer Reserve (CCH Text 5.2)

Name _____   Due Date _____

The *logistic model of population growth* predicts that the population (as a function of time) of deer on the George Deer Reserve in Michigan is given by

$$P(t) = \frac{5.1}{0.03 + e^{-0.59t}}$$

1. Plot the graph of $P(t)$.

2. The *Maximum Sustainable Yield* model for wildlife management says that populations should be maintained at a level where population growth is a maximum. Express this statement in terms of inflection points of $P(t)$.

3. According to the Maximum Sustainable Yield model, at what level should the deer population on the George Deer Reserve be maintained?

4. What is the maximum population of deer that the George Deer Reserve can support? Explain clearly how you arrived at your answer.

Laboratory Exercise 5.4

The Meaning of the Signs of f' and f'' (CCH Text 5.2)

Name _____   Due Date _____

This is a continuation of Lab Exercise 2.5. It requires a color monitor. We will produce a graphical demonstration of the following important principles: If f' is positive, then f is increasing. If f' is negative, then f is decreasing. If f'' is positive, then f is concave up. If f'' is negative, then f is concave down.

First evaluate **Clear[f,g,h,k]** and **f[x_]:=Sin[x]**.

1. Evaluate the following *exactly* as it appears. Be careful to distinguish between square and curly brackets.

$$g := D[f[t],t]$$

$$h[x\_] := If[N[g/.t->x]>=0,f[x],0]$$

$$k[x\_] := If[N[g/.t->x]]<=0,f[x],0]$$

Plot[{h[x],k[x]},{x,-2Pi,2Pi},PlotStyle->{RGBColor[1,0,0],RGBColor[0,1,0]}]

Explain what you see.

2. **D[f[t],t]** is *MATHEMATICA*'s syntax for $f'(t)$, and **N[g/.t->x]** gives a numerical approximation of g at $t = x$. Explain what the formula in Part (1) is doing and how it works.

3. Evaluate the following *exactly* as it appears. Be careful to distinguish between square and curly brackets. (If you want to save time, just change the definition of $g[t]$ and reevaluate the other lines after clearing the definitions.)

```
Clear[g,h,k]
g:= D[f[t],t]
h[x_]:= If[N[g/.t->x]>=0,f[x]]
k[xi_]:= If[N[g/.t->]<=0,f[x]]
Plot[{h[x],k[x]},{x,-2Pi,2Pi},PlotStyle->{RGBColor[1,0,0],RGBColor[0,1,0]}]
```

Explain what you see.

4. D[f[t],t,2] is *MATHEMATICA's* syntax for $f''(t)$. Explain what the formula in Part (3) is doing and how it works.

5. Repeat parts 1-4 above for $f(x) = x^3 - x$.

Plotting families of curves (CCH Text 5.3)

MATHEMATICA's Table function is ideal for plotting families of curves. The syntax is as follows.

Table[*expression*, {*variable, start, end, step size*}]

For example, if evaluate **T= Table[Sin[a*x],{a,1,2,0.5}]**, *MATHEMATICA* will produce a list of expressions of the form $\sin(ax)$ as a ranges from 1 to 2 in steps of 0.5. The result will be the list displayed in Out[1] of Figure 5.4. Hence, if we now evaluate **Plot[Evaluate[T],{x,-2Pi,2Pi}]**, all three graphs will be displayed in order as in Out[2] in Figure 5.4. Use the *MATHEMATICA* function Evaluate to plot a table of functions. See the *MATHEMATICA* manual for more details on this function.

In[1]:=
    T=Table[Sin[a*x],{a,1,2,0.5}]
Out[1]=
    {Sin[x], Sin[1.5 x], Sin[2. x]}
In[2]:=
    Plot[Evaluate[T],{x,-2Pi,2Pi}]

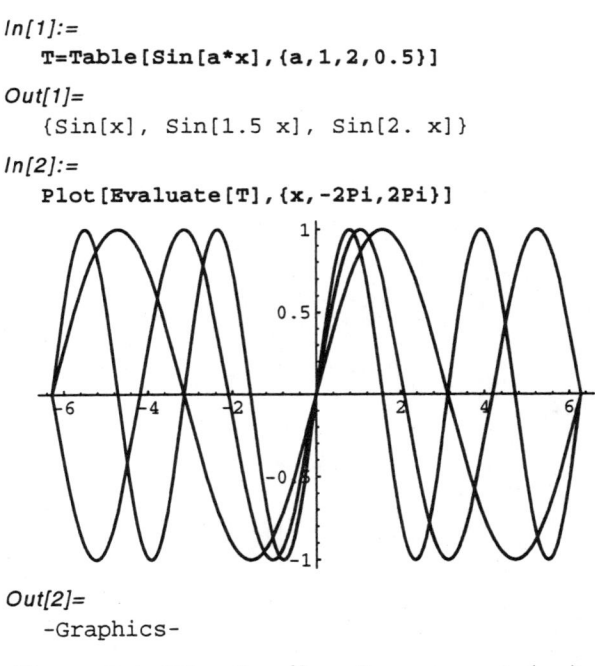

Out[2]=
    -Graphics-

Figure 5.4: **The family of curves,** $\sin(ax)$

Laboratory Exercise 5.5

Families of Curves (CCH Text 5.3)

Name _____   Due Date _____

1. Use the TABLE function as shown in Figure 5.4 to plot $\sin(e^{ax})$ for at least six positive values of a. Clearly label each curve in the printout of your plot with the corresponding value of a.

2. Describe what happens to $\sin(e^{ax})$ for positive x as a gets larger. Describe what happens to $\sin(e^{ax})$ for negative x as a gets larger.

3. Use the TABLE function as shown in Figure 5.4 to plot $x^2 + ax + 1$ for at least five positive values of a and five negative values. Clearly label each curve in the printout of your plot with the corresponding value of a.

4. Describe how the curves $x^2 + ax + 1$ change with a.

5. The vertices of the family of graphs $x^2 + ax + 1$ appear to lie on a single parabola. Find a quadratic function that goes through all their vertices, and plot it along with the family.

Solved Problem 5.4: Welding boxes (CCH Text 5.6)

A company has a contract to build several open metal trash bins. Each has a square base and will hold 1000 cubic feet. The company orders a precut sheet of metal for the bottom of the box, and another that it bends three times to form the four sides. (There is no top.) The company must then weld the seams (one vertical and four horizontal). Records indicate that welding costs $2.10 cents per foot, including labor and materials. The sheet metal costs $1.85 per square foot.

(a) What are the dimensions of the box that costs the least to build?

(b) What is the cost of the box described in Part (a) ?

(c) Another company figures it will make the box 10 feet on a side. How much more does it cost to do this?

Solution to (a): Let h be the height of the box and b the length (and width) of the base:

$$\text{TOTAL COST} = \text{COST OF METAL} + \text{COST OF WELDING}.$$

The bottom has an area of b^2 square feet and there are 4 sides, each having an area of bh square feet. Thus the total cost of the sheet metal used to build the box is $1.85(b^2 + 4bh)$. The total length to be welded is the perimeter of the base, $4b$ feet, plus the height of the seam along a side, h feet. Therefore the welding costs $2.10(4b + h)$. Adding these gives the total cost:

$$\text{COST} = 1.85(4bh + b^2) + 2.10(h + 4b)$$

Since $1000 = hb^2$, we get $h = \frac{1000}{b^2}$. Now we can express the cost in terms of the single variable, b:

$$\text{COST} = 1.85(b^2 + \frac{4000}{b}) + 2.10(4b + \frac{1000}{b^2})$$

Evaluate the cost function as seen in In[2] of Figure 5.5. Use **D[ ]** to differentiate the cost function, and **Simplify** to get the derivative shown in Out[3].

It is best to solve this expression approximately. The plot in Out[8] shows that the derivative crosses the x axis between 10 and 20. Use FindRoot with b initialized at 12. The solution appears as expression 4 of Figure 5.5. We see that the base should be 12.0798 feet on a side, and the height should be $\frac{1000}{12.0798^2} \approx 6.853$ feet.

```
In[1]:=
    cost[b_]:=1.85(b^2+4000/b)+2.10(4b+1000/b^2)
In[2]:=
    D[cost[b],b]
Out[2]=
    2.1 (4 - 2000/b^3) + 1.85 (-4000/b^2 + 2 b)
In[3]:=
    Simplify[%]
Out[3]=
    8.4 - 4200./b^3 - 7400./b^2 + 3.7 b
In[7]:=
    dc[b_]:=%3
In[8]:=
    Plot[dc[b],{b,5,20}]
```

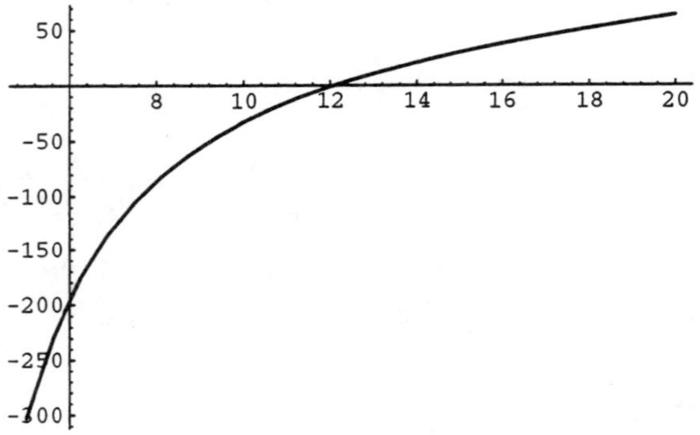

Out[8]=
  -Graphics-

```
In[9]:=
    FindRoot[dc[b]==0,{b,12}]
Out[9]=
    {b -> 12.0798}
In[11]:=
    1000/(12.0798)^2
Out[11]=
    6.853
In[12]:=
    cost[12.0798]
Out[12]=
    998.409
```

Figure 5.5: **Minimum cost of a box**

Solution to (b): We now need to calculate how much the box costs. Evaluate **cost[12.0798]** to plug 12.0797 in for b. The result, $998.409, is Out[12].

Solution to (c): We need to calculate how much the box would cost if we made it 10 feet on each side. Now we evaluate **cost[10]**, and the result (not shown in Figure 5.5) is $1030.00. This is $31.59 (or 3.2%) more than the cheapest way. So the contractor who knows calculus may underbid his or her less astute competitors and win the contract.

Laboratory Exercise 5.6

Building Boxes (CCH Text 5.6)

Name _____   Due Date _____

A contractor wants to bid on an order to make 150,000 boxes out of cardboard that costs 12 cents per square foot. The base of each box must be square and reinforced with an extra layer of cardboard. The contractor must assemble each box by taping the four seams around the bottom, one seam up the side, and one seam on top to make a hinged lid. Company records indicate that taping costs 11 cents per foot, including labor and materials.

1. Assuming that the boxes are to hold three and a half cubic feet, write the cost of a single box as a function of the length of the base.

2. What should the dimensions be to ensure the lowest cost? Show your work.

3. If the contractor wants to make a profit of 17%, what should be the bid?

4. A competitor, who also wants to make a profit of 17%, prepares a bid for boxes that are perfect cubes. By how many dollars does this competitor lose the bid?

Laboratory Exercise 5.7

Building Fuel Tanks (CCH Text 5.6)

Name _____   Due Date _____

A firm is asked to build a cylindrical fuel tank to hold 300 gallons. It orders a rectangular piece of metal which it will roll into a cylinder, and two circular pieces to make the ends. It must then weld the seam of the rolled metal sheet to form a cylinder and the seams to fasten the top and bottom to the cylinder. The rectangular piece must be made of a malleable alloy costing $1.25 per square foot. The circular ends can be a cheaper metal costing 75 cents per square foot. Company records indicate that welding costs 50 cents per foot including labor and materials.

1. Write the cost as a function of the radius of the base.

2. What should the dimensions be to ensure the lowest cost? (There are 7.5 gallons in a cubic foot.) Show your work.

155

3. After the cost is figured, the customer decides she can afford to pay only 80% of the manufacturer's minimum cost computed above. What is the maximum volume that can be produced for this amount?

Laboratory Exercise 5.8

Roads (CCH Text 5.6)

Name _____   Due Date _____

A road from town A to town B must cross a strip of private land and a strip of public land, as shown in Figure 5.6. Due to fees demanded by the owner, the cost of building the road on private land is 20% more per mile than it is on public land. Assuming that the cost on public land is $89,650 per mile, what is the minimum cost of the project? Draw a road map with distances labeled that ensures a minimum cost, and write an explanation that the City Commissioners can understand.

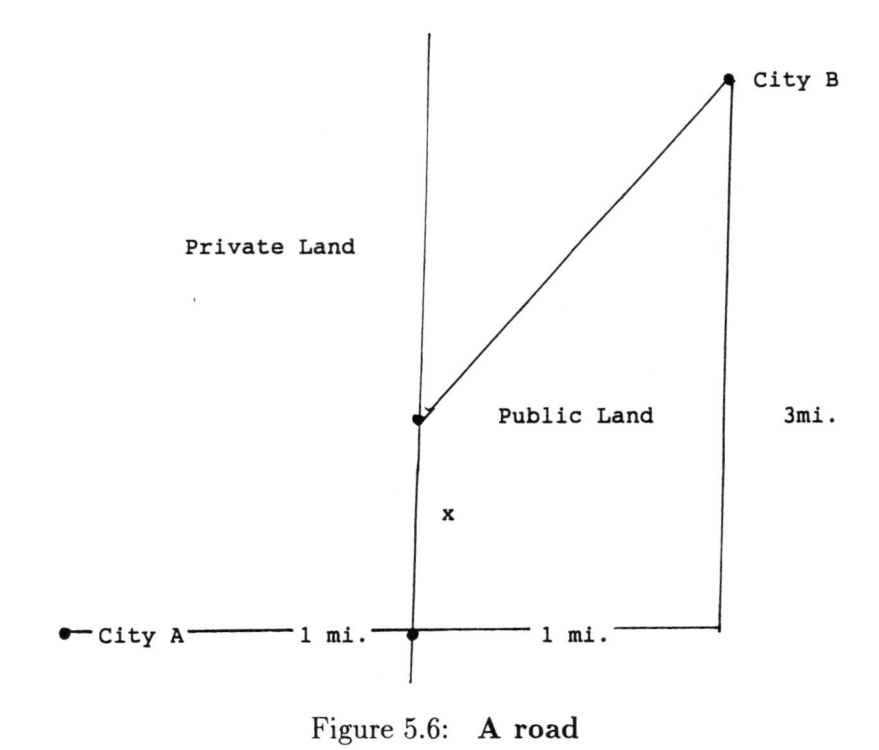

Figure 5.6: **A road**

Solved Problem 5.5: Newton's method (CCH Text 5.7)

Apply 4 steps of Newton's method to $f(x) = \sin x + \cos x$ beginning at $x_0 = 0.6$. Use the package **MYCALC.m** provided in Appendix II to make a picture of this procedure.

Solution: We need to load the package MYCALC.m, which is accomplished by evaluating **<< MYCALC.m**. (This assumes that you have this file in the packages folder of *MATHEMATICA*.)

We must enter f by evaluating **f[x_]:= Sin[x] + Cos[x]**. We want the start point to be $x = 0.6$. Now we can use the function **newt**. Evaluate **newt[f,{x,-5,1.5,4}]**. The syntax of newt can be found in Appendix II. See Out[37] of Figure 5.7.

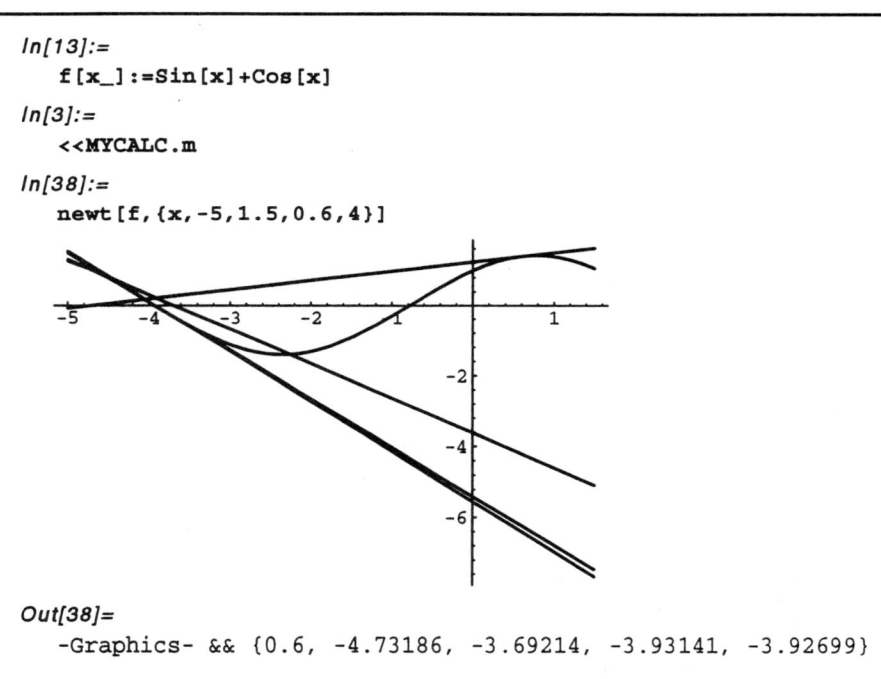

```
In[13]:=
    f[x_]:=Sin[x]+Cos[x]
In[3]:=
    <<MYCALC.m
In[38]:=
    newt[f,{x,-5,1.5,0.6,4}]
```

Out[38]=
-Graphics- && {0.6, -4.73186, -3.69214, -3.93141, -3.92699}

Figure 5.7: **A picture of Newton's method**

Notice that Newton's method did not lead us to the zero closest to the starting point. The last value obtained is -3.92699.

Laboratory Exercise 5.9

Newton's Method (CCH Text 5.7)

Name _____   Due Date _____

1. Apply 5 steps of Newton's method to $x^2 - \cos x$, and use the **MYCALC.m** file to make a picture of the procedure.

2. Make a picture of 10 iterations of Newton's method for $\sin x + \cos x$ for starting points 0.8 and 1.1. Explain what you observe.

3. Find a starting point for Newton's method applied to $\sin x + \cos x$ that will cause Newton's method to get caught in a loop. That is, $x_2 = x_0$. (Hint: If given x_n, find the formula for x_{n+1} in terms of x_n. That is, we want the function $nt(x)$ so that $nt(x_n) = x_{n+1}$. Then we need to find x such that $nt(nt(x)) = nt(x)$. (Why?) In our case, $nt(x) = x - (\sin(x) + \cos(x))/(\cos(x) - \sin(x))$.)

4. Produce a picture of Newton's method applied to $\sin x + \cos x$ using the starting point you found in Part 3.

Chapter 6
Reconstructing a Function from Its Derivative

One of the primary goals of Chapter 6 in your CCH Text is to teach some elementary facts about integration. You may find *MATHEMATICA* useful for checking your answers, but most of the work in this chapter is better done without a computer.

Solved Problem 6.1: Families of antiderivatives (CCH Text 6.3)

Let $f(x) = \ln x$. Plot a family of antiderivatives of f, and find the antiderivative whose value is 3 when $x = 1$.

<u>Solution</u> We evaluate **Clear[f]** and **f[x_]:= Log[x]**. We use *MATHEMATICA*'s Integrate function to find an antiderivative. We evaluate **Integrate[f[x],x]** and the integral appears in Out[3] of Figure 6.1. Thus the family of antiderivatives of $\ln x$ is $x \ln x - x + c$, where c can take any value we choose. However, *MATHEMATICA* does not provide the arbitrary constant c. We must do that ourselves.

We use the TABLE command to plot this family of curves as we did in Laboratory Exercise 5.5. Evaluate **Clear[T]** and **T = Table[x*Log[x]-x + c,{c,-8,8,0.5}]**. This will produce a list of the antiderivatives $c + x \ln x - x$ as c ranges from -8 to 8 in steps of 0.5. Now we evaluate **Plot[Evaluate[T],{x,.5,3}]** to obtain the 17 graphs in Out[6] of Figure 6.1.

To find the antiderivative that has value 3 when $x = 1$, we solve the equation $c + 1 \ln 1 - 1 = 3$ for c (by hand). This gives $c = 4$, so the antiderivative we seek is $4 + x \ln x - x$.

In[1]:=
```
Clear[f]
```
In[2]:=
```
f[x_]:=Log[x]
```
In[3]:=
```
Integrate[f[x],x]
```
Out[3]=
```
-x + x Log[x]
```
In[4]:=
```
Clear[T]
```
In[5]:=
```
T=Table[x*Log[x]-x+c,{c,-8,8,0.5}]
```
Out[5]=
```
{-8 - x + x Log[x], -7.5 - x + x Log[x], -7. - x + x Log[x],
  -6.5 - x + x Log[x], -6. - x + x Log[x],
  -5.5 - x + x Log[x], -5. - x + x Log[x],
  -4.5 - x + x Log[x], -4. - x + x Log[x],
  -3.5 - x + x Log[x], -3. - x + x Log[x],
  -2.5 - x + x Log[x], -2. - x + x Log[x],
  -1.5 - x + x Log[x], -1. - x + x Log[x],
  -0.5 - x + x Log[x], 0. - x + x Log[x],
  0.5 - x + x Log[x], 1. - x + x Log[x], 1.5 - x + x Log[x],
  2. - x + x Log[x], 2.5 - x + x Log[x], 3. - x + x Log[x],
  3.5 - x + x Log[x], 4. - x + x Log[x], 4.5 - x + x Log[x],
  5. - x + x Log[x], 5.5 - x + x Log[x], 6. - x + x Log[x],
  6.5 - x + x Log[x], 7. - x + x Log[x], 7.5 - x + x Log[x],
  8. - x + x Log[x]}
```
In[6]:=
```
Plot[Evaluate[T],{x,.5,3}]
```

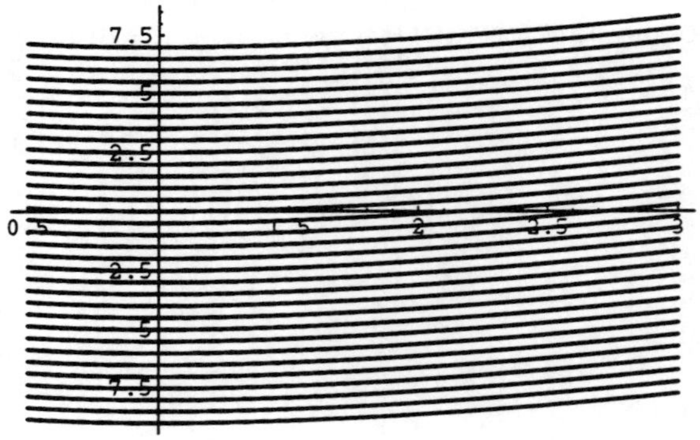

Out[6]=
```
-Graphics-
```

Figure 6.1: **Antiderivatives of** $\ln x$

Laboratory Exercise 6.1

Antiderivatives of arctan x (CCH Text 6.3)

Name _____   Due Date _____

1. Plot a family of at least 10 antiderivatives of arctan x. Explain how the graphs are related.

2. Find an antiderivative of arctan x that has value 5 when $x = 1$ and plot its graph. Can there be more than one such antiderivative? Explain your answer.

Laboratory Exercise 6.2

An Antiderivative of $\sin(x^2)$ (CCH Text 6.3)

Name _____   Due Date _____

Let $f(x) = \sin(x^2)$. When *MATHEMATICA* calculates the antiderivative of $f(x)$, the output contains "FresnelS," which is the integral of $\sin(\pi x^2/2)$! This is no better than what we started with, because there is no "simple" function whose antiderivative is f. However, it does have an antiderivative, and we will examine it.

1. Verify that $g(x) = \displaystyle\int_0^x \sin(x^2)\, dx$ is an antiderivative of f by using *MATHEMATICA* to find the derivative of $g(x)$.

2. Plot the graph of $g(x)$. (*MATHEMATICA* can plot the integral even though it cannot evaluate it exactly. The time required depends on the speed of your machine. It may take a number of minutes!)

3. What is the exact value of $g(0)$? Does any other antiderivative of $f(x)$ have this value at 0? Explain.

4. $g(x)$ is one of many antiderivatives of $f(x)$. Plot the graph of the antiderivative whose value is 3 when $x = 0$. Does any other antiderivative of $f(x)$ have the value 3 at $x = 0$? Explain.

Chapter 7
The Integral

We are familiar with approximating integrals by using left-hand and right-hand Riemann Sums. In this chapter these procedures are referred to as Lhs(n) and Rhs(n), where n is the number of subintervals. Improved approximation schemes are also presented: the *midpoint rule*, MID(n); the *trapezoidal rule*, TRAP(n); and *Simpson's rule*, SIMP(n).

We will help you produce *MATHEMATICA* files (which you can save for later use) that implement these procedures. Rhs(n) and Mid(n) are implemented in Solved Problems 7.2 and 7.3, and the rest are implemented in Lab Exercises 7.2 and 7.3. All five approximation schemes have geometric interpretations that can be represented graphically. The code for this is provided in Appendix II.

MATHEMATICA will also find antiderivatives of a great many functions and thus provides a convenient way to check your answers to exercises. However, *MATHEMATICA* will not tell you *how* it obtains its answers, and it has other limitations.

Solved Problem 7.1: Integrating with *MATHEMATICA* (CCH Text 7.1)

(a) Find $\int \sin \sqrt{x}\, dx$ and check your answer.

(b) Find $\int \sin(x^2)\, dx$.

(c) Approximate $\int_0^1 \sin(x^2)\, dx$.

Solution to (a): We evaluate **Integrate[Sin[Sqrt[x]],x]**, and we see the integral as Out[1] of Figure 7.1.

To check the integral, we differentiate it and see if the result is what we started with. Evaluate **D[%,x]** to get Out[2].

Solution to (b): (This example was the subject of Lab Exercise 6.2.) We evaluate **Integrate[Sin[x^2],x]** and get "FresnelS," which is no simpler than what we want to integrate. This is *MATHEMATICA*'s way of telling us that it cannot find the answer. In this case it's not *MATHEMATICA*'s fault, because $\int \sin(x^2)\, dx$ cannot be expressed in terms of the usual "simple" functions.

Solution to (c): Even though *MATHEMATICA* cannot find the antiderivative, it can still approximate the definite integral. Let's see if *MATHEMATICA* can find the definite integral

```
In[1]:=
    Integrate[Sin[Sqrt[x]],x]
Out[1]=
    -2 (Sqrt[x] Cos[Sqrt[x]] - Sin[Sqrt[x]])
In[2]:=
    D[%,x]
Out[2]=
    Sin[Sqrt[x]]
In[3]:=
    Integrate[Sin[x^2],{x,0,1}]
Out[3]=
    Sqrt[Pi/2] FresnelS[Sqrt[2/Pi]]
In[4]:=
    NIntegrate[Sin[x^2],{x,0,1}]
Out[4]=
    0.310268
```

Figure 7.1: **Integrating with** *MATHEMATICA*

from $x = 0$ to $x = 1$. Evaluate **Integrate[Sin[x^2],{x,0,1}]**. Out[3] of Figure 7.1 does not seem much better. But *MATHEMATICA* has a numerical integration function: NIntegrate. Evaluate **NIntegrate[Sin[x^2],{x,0,1}]**, and the definite integral appears in Out[4].

Laboratory Exercise 7.1

Integration with *MATHEMATICA* (CCH Text 7.1)

Name _____   Due Date _____

1. Use *MATHEMATICA* to find $\int \sec^2 x \, dx$.

2. Based on your answer in Part 1, what should the derivative of $\tan x$ be?

3. Use *MATHEMATICA* to find the derivative of $\tan x$. Reconcile the answer here with your answer in Part 2.

Solved Problem 7.2: Implementing the right-hand rule (CCH Text 7.6)

(a) Create a *MATHEMATICA* function, Rhs(a, b, n), which is the Riemann sum for $\int_a^b f(x)\, dx$ using n subintervals and the right-hand end points of the subintervals.

(b) Use the Rhs function with 20 subintervals to approximate $\int_1^3 x^2\, dx$.

<u>Solution to (a)</u>: In Chapter 3 we did this for the specific function $f(x) = x^2$ and a specific interval. We will just repeat the procedure in general.

The Riemann sum using the right-hand end points of the subintervals is $\sum_{i=1}^{n} f(x_i)\Delta x$, where $\Delta x = \dfrac{b-a}{n}$, and $x_i = a + i\Delta x$.

Evaluate the following input (see Figure 7.2).

Rhs[f_,a_,b_,n_]:= Sum[f[a+i*(b-a)/n](b-a)/n,{i,1,n}]

To use Rhs, define a function $f(x)$ in the usual fashion. Then evaluating Rhs[f,a,b,n] yields the right-hand sum for $f(x)$ from $x = a$ to $x = b$ with n intervals. For example, after entering the definition of Rhs, evaluate **Clear[f], f[x_]:= E^x** and then **Rhs[f,0,2,10]**. The output is the right-hand sum for $y = e^x$ on $(0, 2)$ with 10 intervals.

You may wish to save this in a *MATHEMATICA* file by clicking on the File icon and choosing **Save as**. Give the file a name, for example, RIGHT.MA, and save it.

You can retrieve the definition to use in an active notebook, as follows. First, move the mouse to the File icon and select Open. Find RIGHT.MA and open it, creating a second notebook. You can now copy the function into your active notebook using *MATHEMATICA* editing functions. Hold down the button and drag the mouse over the definition of Rhs, which will highlight the definition. Release the button and the definition remains highlighted. Move the mouse pointer to the Edit icon on the top of the *MATHEMATICA* window, hold down the button, and drag the mouse until Copy is highlighted. Releasing the mouse button copies the highlighted area into buffer. Next, move the mouse to your active notebook and click where you want to copy the function Rhs. Finally, move the mouse back to the Edit icon, depress the button, and drag the mouse to Paste. When it is highlighted, release the mouse button and the definition of Rhs will be copied. It will take some practice to become efficient at this.

<u>Solution to (b)</u>: To use the Rhs function, we evaluate **Clear[f]** and **f[x_]:= x^2**. Now evaluate **Rhs[f,1,3,20]**. To get a numerical approximation, evaluate **N[%]** to get 9.07. The exact value of the integral (calculated by hand) is $\frac{26}{3} \approx 8.66666$. Since x^2 is increasing on the interval $[1, 3]$, the Rhs function produced an overestimate, as expected.

In[1]:=
    rhs[f_,a_,b_,n_]:=
    Sum[f[a+i*(b-a)/n](b-a)/n,{i,1,n}]

In[2]:=
    f[x_]:=x^2

In[3]:=
    rhs[f,1,3,20]

Out[3]=
    $\dfrac{907}{100}$

In[4]:=
    Clear[f]

In[5]:=
    f[x_]:=E^x

In[6]:=
    rhs[f,0,2,10]

Out[6]=
$\dfrac{E^{1/5}}{5} + \dfrac{E^{2/5}}{5} + \dfrac{E^{3/5}}{5} + \dfrac{E^{4/5}}{5} + \dfrac{E}{5} + \dfrac{E^{6/5}}{5} + \dfrac{E^{7/5}}{5} + \dfrac{E^{8/5}}{5} + \dfrac{E^{9/5}}{5} + \dfrac{E^{2}}{5}$

In[7]:=
    N[%]

Out[7]=
    7.04924

Figure 7.2: **Implementing the right-hand rule**

Solved Problem 7.3: Implementing the midpoint rule (CCH Text 7.6)

(a) Create a *MATHEMATICA* function, Mid(f, a, b, n), which is the Riemann sum for $\int_a^b f(x)\,dx$ using n subintervals and the midpoints of the subintervals.

(b) Use the Mid function with 20 subintervals to approximate $\int_1^3 x^2\,dx$.

(c) Use the package **MYCALC.m** in Appendix II to create a picture representing the midpoint sum for 10 subintervals.

Solution to (a): The Riemann sum using the midpoints of the subintervals is $\sum_{i=1}^{n} f(x_i)\Delta x$, where $\Delta x = \dfrac{b-a}{n}$, and $x_i = a + \left(i - \dfrac{1}{2}\right)\Delta x$. (We leave it as an exercise for the reader to show that this formula for x_i gives the midpoint of the ith subinterval.)

The construction here is similar to the one in Solved Problem 7.2. Carefully evaluate the next line (see Figure 7.3):

Mid[f_,a_,b_,n_]:= Sum[f[a+(i -(1/2)(b-a)/n](b-a)/n,{i,1,n}]

```
In[18]:=
    Mid[f_,a_,b_,n_]:= Sum[f[a+(i-1/2)*(b-a)/n](b-a)/n,{i,1,n}]
In[19]:=
    Clear[f]
In[20]:=
    f[x_]:=x^2
In[21]:=
    Mid[f,1,3,20]
Out[21]=
    1733
    ───
    200
In[22]:=
    N[%]
Out[22]=
    8.665
```

Figure 7.3: **Implementing the midpoint rule**

As with the RIGHT.MA file you created in Solved Problem 7.2, you may wish to save this in a file (using the icon File and selecting **Save as**), say MID.MA, or you can have both functions in a single file.

Solution to (b): To use the Mid function, we first evaluate **Clear[f]** and **f[x_]:= x^2** to define f. Next we evaluate **Mid[f,1,3,20]**i, and we get $\frac{1733}{200}$ as in Out[21] of Figure 7.3. Numerically approximating, we obtain 8.665. Notice this answer is much closer to $\frac{26}{3}$, the exact value of the integral, than we got using the Rhs function. Furthermore, x^2 is concave up, so, as expected, Mid gave an underestimate.

Solution to (c): The MYCALC.m package was used extensively in Chapter 3. If you created it at that time, load it using **<< MYCAL.m**. If not, refer to Appendix II.

After the **MYCALC.m**, evaluate **Clear[f]** and **f[x_]:= x^2**. Now evaluate **mid[f,{x,1,3,10}]** to see the picture in Out[4] in Figure 7.4.

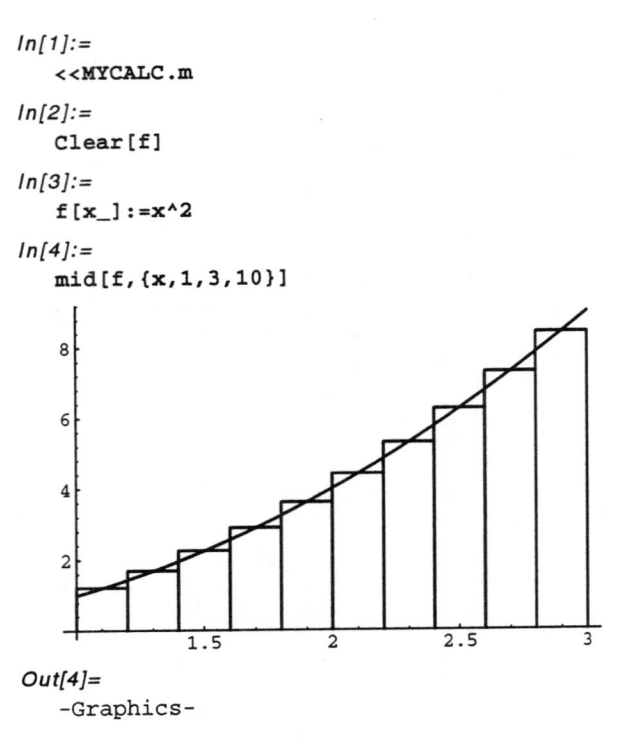

```
In[1]:=
    <<MYCALC.m
In[2]:=
    Clear[f]
In[3]:=
    f[x_]:=x^2
In[4]:=
    mid[f,{x,1,3,10}]
```

Out[4]=
  -Graphics-

Figure 7.4: **Graphical representation of the midpoint rule**

Laboratory Exercise 7.2

Implementing the Left-Hand Rule (CCH Text 7.6)

Name _____   Due Date _____

1. Create a *MATHEMATICA* function, Lhs(f, a, b, n), which is the Riemann sum for $\int_a^b f(x)\, dx$ using n subintervals and the left-hand end points of the subintervals. (You may wish to add this function to a file containing the Rhs and Mid functions. See the instructions at the end of Part (a) of Solved Problem 7.3.)

2. Use the Lhs function with 10 subintervals to approximate $\int_0^{\frac{\pi}{2}} \sin x\, dx$.

3. Use the Rhs function (created in Solved Problem 7.2) with 10 subintervals to approximate $\int_0^{\frac{\pi}{2}} \sin x\, dx$.

4. Discuss how the approximations in Parts 2 and 3 compare to the exact value of $\int_0^{\frac{\pi}{2}} \sin x \, dx$.

5. Use the **MYCALC.m** package given in Appendix II to create pictures representing the left and right sums in Parts 2 and 3. Explain the relationship between the pictures and the approximations.

6. Use the Lhs function with 10 subintervals to approximate $\int_{\frac{\pi}{2}}^{\pi} \sin x \, dx$.

7. Use the Rhs function with 10 subintervals to approximate $\int_{\frac{\pi}{2}}^{\pi} \sin x \, dx$.

8. Discuss how the approximations in Parts 6 and 7 compare to the exact value of $\int_{\frac{\pi}{2}}^{\pi} \sin x \, dx$.

9. Use the **MYCALC.m** package given in Appendix II to create pictures representing the left and right sums in Parts 6 and 7. Explain the relationship between the pictures and the approximations.

10. Discuss the difference between what you observed in Parts 4 and 8. What caused the difference?

Laboratory Exercise 7.3

Implementing the Trapezoidal Rule and Simpson's Rule
(CCH Text 7.7)

Name _____   Due Date _____

1. Create a *MATHEMATICA* function Trap(a, b, n), which is the trapezoidal rule for $\int_a^b f(x)\, dx$ using n subintervals. This is easy if you have saved Lhs and Rhs in a file, as we suggested. We suggest you add Trap to this file. If you need help writing the Trap function in terms of Rhs and Rhs, see page 378 of the CCH Text.

2. Use Trap$(0, \frac{\pi}{2}, 10)$ to find the trapezoidal rule approximation with 10 subintervals for $\int_0^{\frac{\pi}{2}} \sin x\, dx$.

3. Use the MYCALC.m package in Appendix II to create a picture representing the trapezoidal rule in Part 2. Explain the relationship between the picture and the approximation.

4. Create a *MATHEMATICA* function Simp(f, a, b, n) that is the Simpson's rule approximation for $\int_a^b f(x)\, dx$ using n subintervals. This is easy if you have saved Trap and Mid in a file as we suggested. We also suggest you add Simp to this file. If you need help in writing the Simp function in terms of Mid and Trap, see page 387 of the CCH Text. You will find it helpful for Part 7 of this lab to add a function ALL that makes a list of all five.

```
all[f,a,b,n]:={Lhs[f,a,b,n],Rhs[f,a,b,n],Mid[f,a,b,n],Trap[f,a,b,n],Simp[f,a,b,n]}
```

5. Use Simp$(0, \frac{\pi}{2}, 10)$ to approximate $\int_0^{\frac{\pi}{2}} \sin x\, dx$.

6. Use the **MYCALC.m** package from Appendix II to create a picture representing Simpson's rule in Part 5.

7. Discuss how the various approximations (Lhs, Rhs, Mid, Trap, and Simp) compare to the exact value of $\int_0^{\frac{\pi}{2}} \sin x\, dx$. Observe which of the approximations is an overestimate and which is an underestimate. Which approximation is best and which is worst?

Laboratory Exercise 7.4

The Trapezoidal Rule with Error Control (CCH Text 7.7)

Name _____   Due Date _____

A note on the error in using the trapezoidal rule

We noted in Chapter 3 that if f is monotone (increasing or decreasing) on the interval $[a, b]$, then the error in approximating $\int_a^b f(x)\, dx$ using Lhs(n) or Rhs(n) is no more than $|f(b) - f(a)|\frac{(b-a)}{n}$, and that using their average cuts the error in half. Since Trap(n) is in fact the average of Lhs(n) and Rhs(n), we may conclude the following:

If f is a monotone function on $[a, b]$, then the maximum error in using the trapezoidal rule, Trap(n), to approximate $\int_a^b f(x)\, dx$ is less than or equal to

$$|f(b) - f(a)|\frac{b-a}{2n}$$

---

1. What value of n is needed if Trap(n) is to estimate $\int_0^1 \frac{4}{1+x^2}$ with error less than 0.01?

2. Calculate Trap(n) for the value of n you found in Part 1.

3. Find the exact value of $\int_0^1 \dfrac{4}{1+x^2}\, dx$.

4. Obtain *MATHEMATICA*'s 20–place approximation for π. (Use **N[Pi, 20]**.) How close to π is your answer in Part 2?

5. How large would you need to choose n so that Trap(n) gives an approximation that is accurate to 20 decimal places? Discuss the practicality of actually making this calculation.

Solved Problem 7.4: Calculating improper integrals (CCH Text 7.8)

Use *MATHEMATICA* to calculate the following improper integrals. If the integral exists, verify that it is a limit of a Riemann integral.

(a) $\int_0^\infty x^4 e^{-x}\, dx$

(b) $\int_0^2 \dfrac{1}{(1-t)^2}\, dt$

(c) $\int_0^\infty \sin^3 x\, dx$

Solution to (a): We evaluate **NIntegrate[x^4 E^(-x),{x,0,Infinity}]**. The integral appears in Out[1] of Figure 7.5; we obtain the value 24.

```
In[1]:=
   NIntegrate[x^4E^(-x),{x,0,Infinity}]
Out[1]=
   24.
In[2]:=
   Clear[a]
In[3]:=
   a[k_]:=Integrate[x^4E^(-x),{x,0,k}]
In[4]:=
   N[a[10]]
Out[4]=
   23.2979
In[5]:=
   N[a[20]]
Out[5]=
   23.9996
In[6]:=
   Clear[f]
```

```
In[7]:=
   f[t_]:= 1/(1-t)^2
In[8]:=
   Integrate[f[t],{t,0,2}]
   Integrate::idiv: Integral does not converge.
Out[8]=
   Indeterminate
In[9]:=
   Integrate[f[t],{t,0,1}]
   Integrate::idiv: Integral does not converge.
Out[9]=
   Indeterminate
In[10]:=
   Integrate[f[t],t]
Out[10]=
   $-(\dfrac{1}{-1+t})$
```

Figure 7.5: **Evaluating improper integrals**

The definition tells us that $\int_0^\infty x^4 e^{-x}\, dx = \lim_{k\to\infty} \int_0^k x^4 e^{-x}\, dx$. Thus, we evaluate **Clear[a]** and **a[k_]:=Integrate[x^4E^(-x), {x,0,k}]**. Then, evaluating $a(k)$ for larger and larger values of k and approximating (see outputs 4 and 5 in Figure 7.5), we see that $\lim_{k\to\infty}\int_0^k x^4 e^{-x}\, dx = 24$.

Solution to (b): We evaluate **Clear[f]** and **f[t_]:= 1/(1-t)^2**. Now we evaluate **Integrate[f[t],{t,0,2}]**. *MATHEMATICA* responds that the integral does not converge as in Out[8] of Figure 7.5. What's going on?

We notice that the function has a vertical asymptote at $t = 1$ (we leave it to the reader to plot it), so this is an improper integral and must be checked for convergence. Our integral exists, provided *both* the integrals $\int_0^1 \frac{1}{(1-t)^2}\, dt$ and $\int_1^2 \frac{1}{(1-t)^2}\, dt$ converge. Evaluating **Integrate[f[t],{t,0,1}]** we again see (Out[9]) that the integral does not converge; that is, it diverges. Hence $\int_0^2 \frac{1}{(1-t)^2}$ diverges. (You should be able to verify this by direct calculation without a computer.)

If we evaluate **Integrate[f[t],t]** we get $-\frac{1}{-1+t}$. One might be misled to think then that the integral $-\frac{1}{-1+2} - (-\frac{1}{-1+0}) = -2$, which, of course, is wrong. Once again the moral is clear: Computer output must always be viewed critically.

Solution to (c): We evaluate **NIntegrate[Sin[x]^3,{x,0, Infinity}]** and get Out[1] of Figure 7.6. We might conjecture what *MATHEMATICA* had trouble doing, but it makes little difference – our job is to figure out what the integral is. Let's evaluate **Integrate[Sin[x] ^3, {x, 0, k}]**. If we ask *MATHEMATICA* to calculate the limit of Out[2] as k goes to infinity, it will return the limit unevaluated. (Try it.)

In[1]:=
NIntegrate[Sin[x]^3,{x,0,Infinity}]

NIntegrate::slwcon:
  Numerical integration converging too slowly; suspect one
    of the following: singularity, oscillatory integrand, or
    insufficient WorkingPrecision.

NIntegrate::ncvb:
  NIntegrate failed to converge to prescribed accuracy after
    7 recursive bisections in x near x = 2.28833 10^{56}.

Out[1]=
  $-3.30763 \ 10^{3496}$

In[2]:=
Integrate[Sin[x]^3,{x,0,k}]

Out[2]=
  $\frac{2}{3} + \frac{-9 \cos[k] + \cos[3\ k]}{12}$

In[3]:=
Limit[%,k->Infinity]

Out[3]=
  $\text{Limit}[\frac{2}{3} + \frac{-9 \cos[k] + \cos[3\ k]}{12}, \ k \ \text{->} \ \text{Infinity}]$

In[4]:=
Plot[2/3+(-9Cos[k]+Cos[3k])/12,{k,10,40}]

Out[4]=
-Graphics-

Figure 7.6: **Divergence of** $\int_0^\infty \sin^3 x \ dx$

We will plot the function. Evaluate **Plot[2/3 + (-9Cos[x] +Cos[3x])/12,{x,10,40}]** . From the graph in Out[4] of Figure 7.6 it appears that the antiderivative wiggles back and forth between 0 and about 1.25 and does not approach a single value for large k. We conclude that the limit does not exist, hence $\int_0^\infty \sin^3 x \ dx$ diverges.

187

Laboratory Exercise 7.5

Evaluating Improper Integrals (CCH Text 7.8)

Name _____   Due Date _____

Calculate each of the following integrals. Verify your answers by calculating the appropriate limit. Include any graphs you may have used to help you.

1. $\int_0^\infty x^3 2^{-x}\, dx$

2. $\int_0^1 \ln x\, dx$

3. $\int_0^4 \frac{1}{x^2 - 2}\, dx$

4. $\int_0^\infty \cos^5 x\, dx$

5. $\displaystyle\int_1^\infty \frac{1}{\arctan x}\, dx$

6. $\displaystyle\int_2^\infty \frac{1}{t^2+3t-4}\, dt$

7. $\displaystyle\int_1^2 \frac{1}{t^2+3t-4}\, dt$

8. $\displaystyle\int_1^\infty \frac{1}{t^2+3t-4}\, dt$

Laboratory Exercise 7.6

The Gamma Function (CCH Text 7.8)

Name _____   Due Date _____

We encountered the **Gamma** function in Lab Exercise 2.5, but we were not told how it was defined. It is defined by $\Gamma(x) = \int_0^\infty t^{x-1} e^{-t}\, dt$ and is important in several areas of mathematics.

1. Use *MATHEMATICA* to integrate $t^{x-1}e^{-t}$ on $[0, \infty]$. Evaluate **h[t_,x_]:= t^(x-1)E^(-t)** and **G[x_]:= NIntegrate[t^(x-1)E^(-t),{t,0,Infinity}]**.

2. Find $G(\frac{1}{2})$.

3. Find $\Gamma(\frac{1}{2})$. Recall, that you can use *MATHEMATICA*'s Gamma function.

4. Compare the accuracy of the approximations in Parts 2 and 3, and explain the reason for what you observed.

5. In Lab Exercise 2.5 we estimated $\Gamma'(1)$ in different ways. Here we will try another.
   (a) Find $\dfrac{d}{dx} t^{x-1} e^{-t}$.
   (b) Substitute $x = 1$ into the result in (a).
   (c) Numerically integrate the result found in (b) from 0 to ∞, and compare it with the estimates you obtained in Lab Exercise 2.5. <u>Note:</u> What you just did is called "differentiating under the integral sign." It often works but must be justified.

Solved Problem 7.5: Approximating improper integrals (CCH Text 7.9)

Determine whether each of the integrals converges or diverges. If it converges, find or estimate its value.

(a) $\int_4^\infty \frac{1}{\ln t}\, dt$

(b) $\int_1^\infty \frac{1}{t^5 + t + 3}\, dt$

Solution to (a): When we evaluate **Integrate[1/Log[x],{x,4,Infinity}]**, we get some error message and the integral back. Trying Integrate[1/Log[x],x] does not yield much more success. Next, let's try to numerically approximate the integral by evaluating **NIntegrate[1/Log[x], {x, 4, Infinity}]**. As in Out[1] of Figure 7.7, we get some error messages and an answer of 1.5422110^{3494}! *MATHEMATICA* tells us that the NIntegral failed to converge. We guess that the integral diverges.

How can we be sure that it does not converge? If we evaluate **Plot[1/Log[x], {x,4,20}]** we see (Out[2]) that the graph looks like the graph of $y = 1/x$. If we evaluate **Plot[{1/Log[x],1/x},{x,4,20}]** as in Out[3] in Figure 7.7 we see that $1/x < \ln x$ on this interval. Try plotting the graphs with domain $[20, 200]$. This provides further evidence for the inequality $\frac{1}{\ln t} > \frac{1}{t}$ for $t > 4$. (You should be able to show that this is always true. How does $\ln t$ compare with t?)

We conclude that the two integrals are related as follows:

$$\int_4^\infty \frac{1}{\ln t}\, dt \geq \int_4^\infty \frac{1}{t}\, dt$$

But $\int_4^\infty \frac{1}{t}\, dt$ diverges. Thus, by the Comparison Test, $\int_4^\infty \frac{1}{\ln t}\, dt$ must also diverge.

You should not be disappointed if *MATHEMATICA* cannot provide an immediate answer to an improper integral. This is not a weakness of *MATHEMATICA*, but a strength. The message that NIntegrate failed to converge alerts us to be suspicious of the output. It is inherently impossible for a computer program to determine convergence numerically because it is only sampling a finite number of points, so *we* must decide this. We would have been wiser to check out convergence on our own *before* asking *MATHEMATICA* for an approximation of an improper integral in this case.

Solution to (b): Evaluate **Integrate[1/(x^5+t+3),{t,1,Infinity}]**, and again *MATHEMATICA* returns the input, indicating that it cannot calculate the integral. (Try it.)

We'll apply the lesson we learned from Part (a) by checking for convergence before asking for an approximation. For $t \geq 0$, $t^5 + t + 3 > t^5 > 0$, so $\frac{1}{t^5 + t + 3} < \frac{1}{t^5}$.

In[1]:=
NIntegrate[1/Log[x],{x,4,Infinity}]

NIntegrate::slwcon:
  Numerical integration converging too slowly; suspect one
    of the following: singularity, oscillatory integrand, or
    insufficient WorkingPrecision.

NIntegrate::ncvb:
  NIntegrate failed to converge to prescribed accuracy after
    7 recursive bisections in x near x = 2.28833 10^{56}.

Out[1]=
  1.54221 10^{3494}

In[2]:=
Plot[1/Log[x],{x,4,20}]

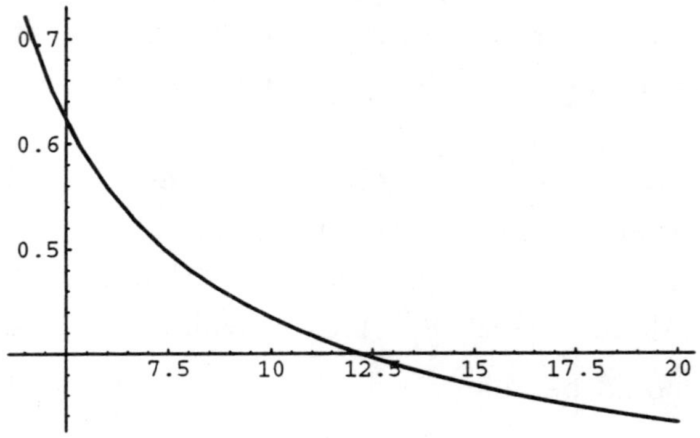

Out[2]=
  -Graphics-

In[3]:=
Plot[{1/Log[x],1/x},{x,4,20}]

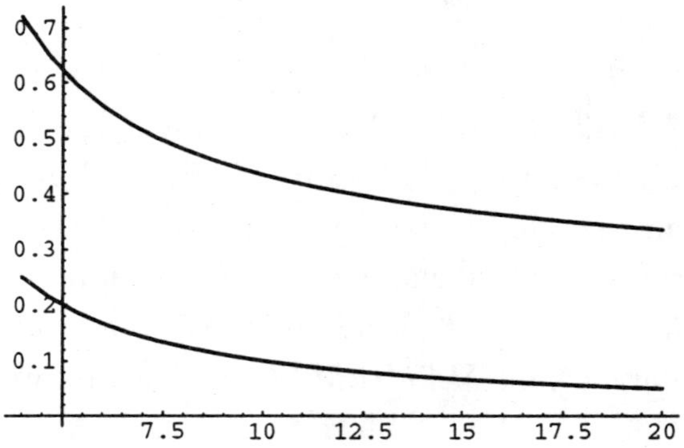

Out[3]=
  -Graphics-

Figure 7.7: $\int_4^\infty \dfrac{1}{\ln t}\, dt$ and $\dfrac{1}{\ln t}$ compared with $\dfrac{1}{t}$

194

Hence, $\int_1^\infty \frac{1}{t^5+t+3}\,dt \leq \int_1^\infty \frac{1}{t^5}\,dt$. We know that the latter integral converges. Therefore, by the Comparison Test, $\int_1^\infty \frac{1}{t^5+t+3}\,dt$ must also converge.

We can now feel comfortable asking for an approximation of the integral. If we evaluate **NIntegrate[1/(t^5+t+3),{t,1,Infinity}]**, we obtain the value 0.10874 in Out[1] of Figure 7.8. We did not see any warnings, so we may feel pretty good about the answer. But how accurate is it? See Solved Problem 7.6.

In[1]:=
    NIntegrate[1/(t^5+t+3),{t,1,Infinity}]

Out[1]=
    0.10874

Figure 7.8: **Approximating** $\int_1^\infty \frac{1}{t^5+t+3}\,dt$

Laboratory Exercise 7.7

Approximating Improper Integrals (CCH Text 7.9)

Name _____   Due Date _____

MATHEMATICA cannot evaluate the following integrals exactly. Determine if each integral converges. (A suggested comparison is given for each. You should analyze the comparison both graphically and algebraically.) For those integrals that converge, report *MATHEMATICA*'s approximation.

1. $\int_2^\infty \dfrac{1}{x^2 \ln x}\, dx$ (Suggested comparison: $\int_2^\infty \dfrac{1}{x^2}\, dx$)

2. $\int_2^\infty \dfrac{1}{\ln^2 t}\, dt$ (Suggested comparison: $\int_2^\infty \dfrac{1}{t \ln t}\, dt$)

3. $\int_0^\infty \dfrac{1}{x+e^x}\, dx$ (Suggested comparison: $\int_0^\infty e^{-x}\, dx$)

4. $\int_0^1 \dfrac{1}{x^3+\sqrt{x}}\, dx$ (Suggested comparison: $\int_0^1 x^{-\frac{1}{2}}\, dx$)

Solved Problem 7.6: Approximating improper integrals II (CCH Text 7.9)

In Solved Problem 7.5 we showed that $\int_1^\infty \frac{1}{t^5+t+3} \, dt$ converges, and we approximated it. The goal here is to get an approximation to a *known degree of accuracy*.

(a) Find a value of k so that $\int_1^k \frac{1}{t^5+t+3} \, dt$ approximates $\int_1^\infty \frac{1}{t^5+t+3} \, dt$ to two decimal places.

(b) Find an approximation to $\int_1^\infty \frac{1}{t^5+t+3} \, dt$ that you know is accurate to one decimal place.

<u>Solution to (a)</u>: We will not use a computer here. For $\int_1^k \frac{1}{t^5+t+3} \, dt$ to approximate $\int_1^\infty \frac{1}{t^5+t+3} \, dt$ to two decimal places, we must have

$$\int_1^\infty \frac{1}{t^5+t+3} \, dt - \int_1^k \frac{1}{t^5+t+3} \, dt < 0.005$$

But $\int_1^\infty \frac{1}{t^5+t+3} \, dt - \int_1^k \frac{1}{t^5+t+3} \, dt = \int_k^\infty \frac{1}{t^5+t+3} \, dt$. Thus we need

$$\int_k^\infty \frac{1}{t^5+t+3} \, dt < 0.005$$

Using the fact that $\frac{1}{t^5+t+3} < \frac{1}{t^5}$, we can conclude that for any $k > 0$

$$\int_k^\infty \frac{1}{t^5+t+3} \, dt < \int_k^\infty \frac{1}{t^5} \, dt = \frac{1}{4k^4}$$

Therefore, we want $\frac{1}{4k^4} < 0.005$ or $k > \frac{1}{\sqrt[4]{4(0.005)}} \approx 2.65914$.

We conclude that $\int_1^3 \frac{1}{t^5+t+3} \, dt$ approximates $\int_1^\infty \frac{1}{t^5+t+3} \, dt$ to within two decimal places.

<u>Solution to (b)</u>: From Part (a), we know that $\int_1^3 \frac{1}{t^5+t+3} \, dt$ approximates $\int_1^\infty \frac{1}{t^5+t+3} \, dt$ to within two decimal places, but *MATHEMATICA* can't give us the exact answer to this integral either. (Try it.)

We ask *MATHEMATICA* to approximate the answer using NIntegrate for $\int_1^3 \frac{1}{t^5+t+3} dt$. The result, 0.105689 in Out[1] of Figure 7.8, is very close to the approximation, 0.10874, that *MATHEMATICA* gave us for $\int_1^\infty \frac{1}{t^5+t+3} dt$ in Out[2] in Figure 7.8. This gives us confidence that *MATHEMATICA*'s approximations are accurate, but how can we be sure?

We use Trap$(f, 1, 3, n)$ to approximate $\int_1^3 \frac{1}{t^5+t+3} dt$ so that we *know* what the error is. To ensure two-place accuracy, we need to have

$$|f(3) - f(1)|\frac{3-1}{2n} \leq 0.005$$

This gives $n > 39.1967$, so we will use Trap$(f, 1, 3, 40)$. With the file containing our Trap function loaded, we evaluate the input line with the definition of Trap(f,a,b,n). Then we evaluate **Clear[f], f[x_]:= 1/(x^5+x+3)**. Then evaluate **Trap[f,1,3,40]**, as seen in outputs 5 and 6 of Figure 7.9.

```
In[3]:=
    f[t_]:=1/(t^5+t+3)
In[5]:=
    Trap[f,1,3,40]
Out[5]=
    19656577852163430326347396554119113156089397900723308827184 6\
    074990070978323713539218119185084183459751752291596710956\
    760757129738122620232441942328360294601132924537325489827\
    732040470991564377266918907765213611137104704280158984574\
    1408 / 185899130500438668698566709692836851469681287956 12\
    713772910261622365520953212826869195677401911774234493 68\
    517937760096976212014699426187478048037436408137370544696\
    888958022292500852271663345204705661683360825933191550352\
    367976941773875
In[6]:=
    N[%]
Out[6]=
    0.105738
```

Figure 7.9: **Approximation and accuracy**

We *know* now that 0.105737 approximates $\int_1^3 \frac{1}{t^5+t+3} dt$ to two decimal places, and that $\int_1^3 \frac{1}{t^5+t+3} dt$ approximates $\int_1^\infty \frac{1}{t^5+t+3} dt$ to two decimal places. Thus we can be sure that 0.105737 approximates $\int_1^\infty \frac{1}{t^5+t+3} dt$ to one decimal place. (Explain why.)

200

Why all the fuss? *MATHEMATICA's* original approximation, 0.10874, to $\int_1^\infty \frac{1}{t^5+t+1}\,dt$ turned out to be quite good. Most of *MATHEMATICA's* approximations are. In general, if you do not see the warnings, you can be fairly confident of the numerical approximations of integrals that *MATHEMATICA* gives. But to demonstrate a *definite* level of accuracy, you need a procedure such as we used here.

Laboratory Exercise 7.8

Approximating Improper Integrals II (CCH Text 7.9)

Name _____   Due Date _____

1. Show that $\int_1^\infty \dfrac{1}{t^6+t+1}\,dt$ converges.

2. Find a value of k so that $\int_1^k \dfrac{1}{t^6+t+1}\,dt$ approximates $\int_1^\infty \dfrac{1}{t^6+t+1}\,dt$ to two decimal places. Explain.

3. Find an approximation of $\int_1^\infty \dfrac{1}{t^6+t+1}\,dt$ that you are certain is accurate to one decimal place. Explain.

Chapter 8
Using the Definite Integral

This chapter looks at applications of the definite integral to geometry and physics. *MATHEMATICA* can perform the calculations, but it's up to you to set up the problems.

Solved Problem 8.1: An oil slick (CCH Text 8.1)

Oil has spilled into a straight river 10 meters wide and has drifted downstream. The density of the oil slick is given by $d(x) = \dfrac{100}{1+x+x^2}$ kilograms per square meter, where $x \geq 0$ is the number of meters downstream from the source of the spill. (See Figure 8.1.) We assume that the density of the slick does not vary from one shore to the other.

(a) How much oil is within 25 meters of the source of the spill?

(b) How much oil was spilled into the river?

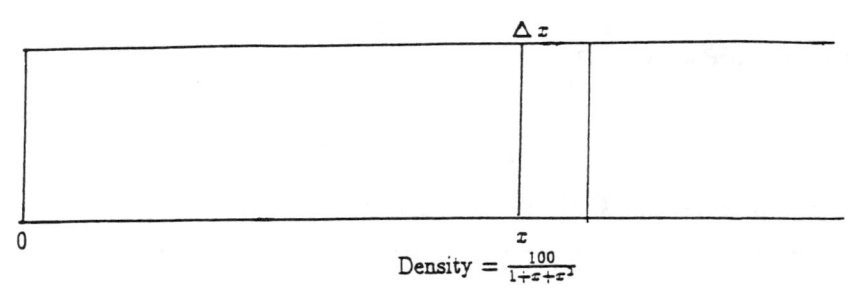

Figure 8.1: **An oil spill in a straight river**

Solution to (a): Since the density of the slick does not vary from one shore to the other, the amount of oil in a slice of the river of width Δx is

$$\text{Density} \times \text{Area} = d(x)10\Delta x \quad \text{(See Figure 8.1.)}$$

Thus the mass of oil in a stretch of the river is approximated by a Riemann sum of the form $\sum d(x_i) 10\Delta x$.

In the 25 meters downstream from the source, the mass is given by $\int_0^{25} \frac{1000}{1+x+x^2}\,dx$.

We evaluate **Integrate[1000/(1+x+x^2),{x,0,25}]**. The integral appears in Out[1] of Figure 8.2. We evaluate **N[%]** to get 1170 kilograms as seen in Out[2].

```
In[1]:=
   Integrate[1000/(1+x+x^2),{x,0,25}]
Out[1]=
   -1000 Pi     2000 ArcTan[17 Sqrt[3]]
   --------  +  -----------------------
   3 Sqrt[3]            Sqrt[3]
In[2]:=
   N[%]
Out[2]=
   1170.
In[3]:=
   Integrate[1000/(1+x+x^2),{x,0,Infinity}]
Out[3]=
   2000 Pi
   --------
   3 Sqrt[3]
```

```
In[4]:=
   N[%]
Out[4]=
   1209.2
In[5]:=
   Integrate[1000/(1+x+x^2),{x,0,10000}]
Out[5]=
   -1000 Pi     2000 ArcTan[6667 Sqrt[3]]
   --------  +  -------------------------
   3 Sqrt[3]             Sqrt[3]
In[6]:=
   N[%]
Out[6]=
   1209.1
```

Figure 8.2: **Calculating the mass of an oil spill**

Solution to (b): The total oil spilled is the total mass of oil downstream. We weren't given the extent of the spill, so we will calculate the total amount of oil that would be in a slick of infinite length that is, $\int_0^\infty \frac{1000}{1+x+x^2}\,dx$. This integral is Out[3] of Figure 8.2, and its approximate value, 1209.2 kilograms, is Out[4].

In Out[5] and Out[6] we have calculated the total mass of oil in a 10,000–meter slick. There is less than one kilogram more in a river of infinite length. Therefore, it makes sense to assume the slick is infinite in length.

Solved Problem 8.2: Area and center of mass (CCH Text 8.1)

A thin sheet of metal is shaped like the region in the first quadrant between $y = \sin x$ and $y = \frac{x}{2}$. (See the region in Out[1] of Figure 8.3.)

(a) Find the area of the region.

(b) Find the x coordinate of the center of mass of the region.

In[1]:=
   Plot[{Sin[x],x/2},{x,0,Pi}]

Out[1]=
   -Graphics-

In[3]:=
   FindRoot[Sin[x]==x/2,{x,1.8}]

Out[3]=
   {x -> 1.89549}

In[4]:=
   Integrate[Sin[x]-x/2,{x,0,1.89549}]

Out[4]=
   0.420798

In[5]:=
   Integrate[x(Sin[x]-x/2),{x,0,1.89549}]

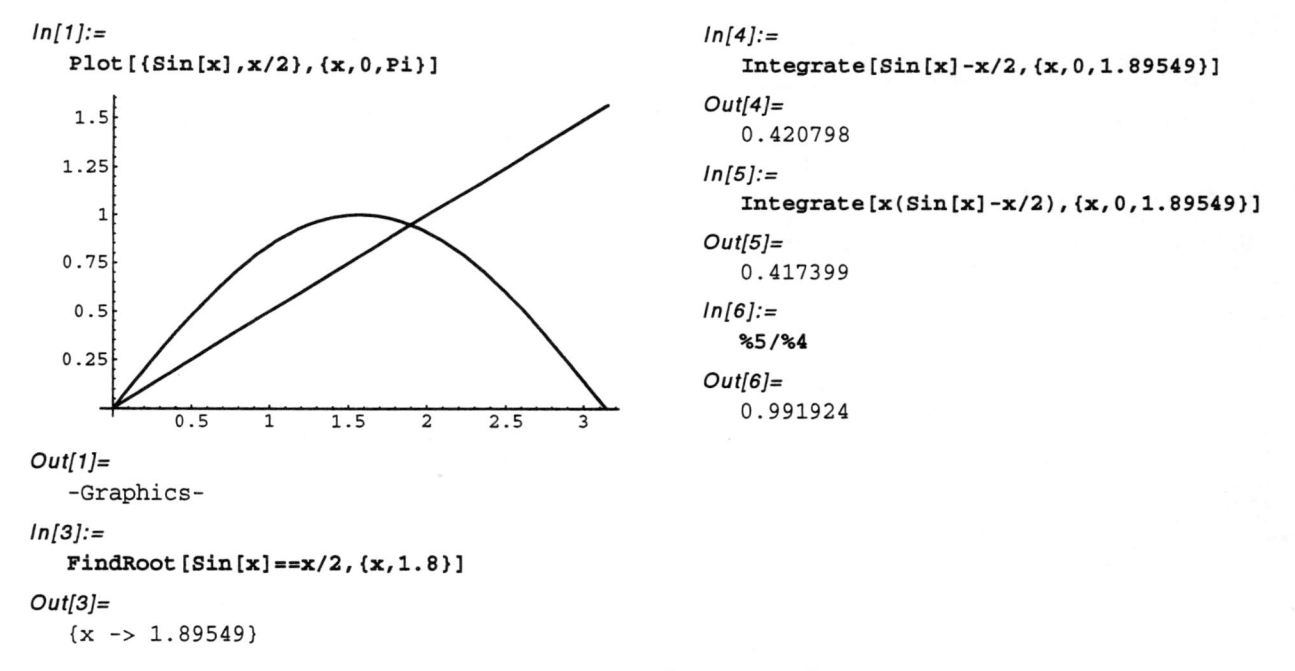

Out[5]=
   0.417399

In[6]:=
   %5/%4

Out[6]=
   0.991924

Figure 8.3: **Area and center of mass**

Solution to (a) We evaluate **Plot[{Sin[x],x/2},{x,0,Pi}]**, as seen in Out[1] of Figure 8.3. From the graph we see that to find the area we will have to integrate $\sin x - \frac{x}{2}$ from $x = 0$ to the point where the graphs cross to the right of $x = 0$. Therefore, we must solve $\sin x = \frac{x}{2}$. This cannot be done exactly, so we use **FindRoot[Sin[x]==x/2,{x,1.8}]**. (We refer the reader to Section 8 of Appendix I for details.) The solution is seen in Out[3]. Thus, the area is
$$\int_0^{1.89549} \sin x - \frac{x}{2}\, dx.$$

207

Evaluate **Integrate[Sin[x]-x/2,{x,0,1.89549}]** to find the area, 0.420798, as Out[4] of Figure 8.3.

Solution to (b): The x coordinate of the center of mass of the region bounded above by $y = f(x)$ and below by $y = g(x)$ on the interval $[a, b]$ is given by the formula

$$\frac{\int_a^b x(f(x) - g(x))\, dx}{\int_a^b f(x) - g(x)\, dx}$$

The numerator is called the "first moment" of the region about the y axis. The denominator is the area. If we placed this region on a thin rod parallel to the y axis at this x value, the region would balance on the rod. (See problem 2 in Section 8.1 of the CCH Text for a related exercise.)

To finish this problem, we need to find $\int_0^{1.89549} x(\sin x - \frac{x}{2})\, dx$ and divide it by the area we found in Part (a).

Thus, we evaluate **Integrate[x(Sin[x]-x/2),{x,0,1.89549}]**, and obtain the moment, 0.417399, as Out[5] in Figure 8.3. The ratio, 0.991924, appears as Out[6]. If the line, $x = 0.991924$, were a thin rod, the region would balance on the rod.

Laboratory Exercise 8.1

An Oil Spill in the Ocean (CCH Text 8.1)

Name _____   Due Date _____

A tanker ship illegally flushes its bilges off the Port of Galveston, making a circular oil slick with the ship at its center. The density of the slick x miles from the tanker is $\dfrac{8000}{1+e^x}$ gallons per square mile.

1. Set up a Riemann sum that approximates the total amount of oil in the slick when its radius is 8 miles. (<u>Hint</u>: Example 5 in section 8.1 of the CCH Text offers a useful model for this problem.)

2. Find the amount of oil in the 8-mile slick.

3. Change the limits on your integral in Part 2 to 0 and ∞. Evaluate this new integral and explain in practical terms what this number means.

Laboratory Exercise 8.2

The Center of Mass of a Sculpture (CCH Text 8.1)

Name _____   Due Date _____

   An art student was given a circular metal disk two feet in diameter and told to drill a small hole in it so that when the disk is cut in half and the piece with the hole is placed atop a spike stuck in the hole, it will balance. Not knowing about integrals, the artist drilled a hole at a point half-way between the center and the edge.

1. Where *should* the art student have drilled the hole? (<u>Hint</u>: Consider the right half of the circle $x^2 + y^2 = 1$, and see Solved Problem 8.2 for a discussion of computing a center of mass.)

2. Now that he has made the mistake, he decides that rather than drill a second hole, he will cut the piece with the hole in it in such a way that it will balance on the spike at the point of the hole. Explain clearly how the disk should be cut so that our artist friend can understand.

3. What is the area of the piece of the metal disk that balances at the point where the hole $(\frac{1}{2}, 0)$ was drilled?

Solved Problem 8.3: Arc length, area, and volume (CCH Text 8.2)

Let $f(x) = \dfrac{e^x + e^{-x}}{2}$ and $g(x) = e^x + e^{-x}$.

(a) Find the lengths of the curves, f and g, from $x = 0$ to $x = 1$, and compare the answers.

(b) Find the area bounded by the graph of f and the x axis from $x = 0$ to $x = 1$.

(c) Find the volume of the solid generated if the region in Part (b) is rotated around the x axis.

<u>Solution to (a)</u>: First, we evaluate **Clear[f,g,df,dg]**, **f[x_]:= (E^x+E^(-x))/2**, and **g[x_]:= 2f[x]**, as in inputs 1-3 in Figure 8.4. Next evaluate **df[x_] := D[f[x],x]** and **dg[x_]:= 2df[x]**.

Next, we evaluate **Integrate[Sqrt[1 + df[x]^2], {x,0,1}]** to get the exact value in Out[8]. Then evaluate **N[%]** to get the approximation 1.1752.

We follow the same procedure for g in In[15]. This time when we evaluate **Integrate[Sqrt[1 + dg[x]^ 2], { x, 0, 1 }]** *MATHEMATICA* returns some messages, begins calculating, but cannot produce an exact answer. (It is rare that an arc length integral can be evaluated exactly.) We evaluate using the NIntegrate function to get the decimal approximation 1.55189 in Out[14].

The second function is two times the first, so one might expect its arc length to be twice as great. As we see, that is not the case.

<u>Solution to (b)</u>: The area is $\int_0^1 (e^x + e^{-x})/2 \; dx$. (The reader should plot a picture of the region.) We evaluate **Integrate[f[x],{x,0,1}]**. The exact value of the area and its decimal approximation appear in Out[16] and Out[17] of Figure 8.4

<u>Solution to (c)</u>: We continue assuming the definition $f[x] := (e^x + e^{-x})/2$ remains in effect. Slice the solid into "disks" with their centers on the x axis. Each disk has volume $\pi y^2 \Delta x$. Thus, the total volume is approximated by a Riemann sum of the form $\sum \pi y^2 \Delta x$. Therefore the volume is given by the integral $\int_0^1 \pi y^2 \; dx$, which appears in In[18]. We get $(-1+4e^2+e^4)\pi/8e^2 \approx 4.41933$ (not shown).

```
In[1]:=
   Clear[f,g,df,dg]
In[2]:=
   f[x_]:=(E^x+E^(-x))/2
In[3]:=
   g[x_]:=2f[x]
In[4]:=
   df[x_]:=D[f[x],x]
In[5]:=
   dg[x_]:=2df[x]
In[13]:=
   dg[x]
Out[13]=
   -E^-x + E^x
In[8]:=
   Integrate[Sqrt[1+df[x]^2],{x,0,1}]
Out[8]=
```
$$\frac{(-1 + E^2)\ \text{Sqrt}[2 + E^{-2} + E^2]}{2\ (1 + E^2)}$$
```
In[9]:=
   N[%]
Out[9]=
   1.1752
```

```
In[15]:=
   Integrate[Sqrt[1+(E^x-E^(-x))^2],{x,0,1}]
   On::none: Message SeriesData::csa not found.
   On::none: Message SeriesData::csa not found.
   On::none: Message SeriesData::csa not found.
   General::stop:
      Further output of On::none
         will be suppressed during this calculation.
Out[15]=
   $Aborted
In[14]:=
   NIntegrate[Sqrt[1+(E^x-E^(-x))^2],{x,0,1}]
Out[14]=
   1.55189
In[16]:=
   Integrate[f[x],{x,0,1}]
Out[16]=
```
$$\frac{-1}{2\ E} + \frac{E}{2}$$
```
In[17]:=
   N[%]
Out[17]=
   1.1752
In[18]:=
   Integrate[Pi*f[x]^2,{x,0,1}]
Out[18]=
```
$$\frac{(-1 + 4\ E^2 + E^4)\ \text{Pi}}{8\ E^2}$$

Figure 8.4: **Arc length, area, and volume**

Laboratory Exercise 8.3

Arc Length and Volume (CCH Text 8.2)

Name _____   Due Date _____

In each of the following, find the length of the given arc and the volume of the solid obtained by rotating the given region, R, about the x axis.

1. Arc: $y = \sin x$ from $x = 0$ to $x = \pi$.

   R: One arch of the sine curve above the x axis.

2. Arc: $y = x^2$ from $x = 0$ to $x = 1$.

   R: The region between $y = x^2$ and $y = \sin x$ in the first quadrant.

3. Arc: $y = \sqrt{1 - \frac{x^2}{9}}$ from $x = 0$ to $x = 3$. (The integral is a special case of an *elliptic integral*.)

   R: The region in the first quadrant that lies beneath $y = \sqrt{1 - \frac{x^2}{9}}$ and outside the unit circle.

Laboratory Exercise 8.4

Arc Length and Limits (CCH Text 8.2)

Name _____   Due Date _____

In this problem we examine the length of the arc of the curve $y = x^n$ on the interval $[0, 1]$ for different values of n.

1. Approximate the length of the arc of the curve $y = x^n$ on the interval $[0, 1]$ for $n = 1, 10, 20,$ and 100.

2. For the case $n = 1$ explain how you can get the answer very quickly by just looking at the graph.

3. Discuss any pattern or trend you see in the calculations in Part 1.

4. Plot the graphs of the four curves in Part 1, and use them to help explain what is happening to the arc lengths as n gets larger.

5. Based on all the above, find $\lim_{n\to\infty} \int_0^1 \sqrt{1+(n+1)^2 x^{2n}}\, dx$.

    Hint: $\sqrt{1+(n+1)^2 x^{2n}} = \sqrt{1+\left(\dfrac{d}{dx}x^{n+1}\right)^2}$.

6. Repeat Parts 1 through 4 using the curve $y = \sqrt{1-x^2}$ on the interval $[-1, 1]$.

Laboratory Exercise 8.5

Ratio of Arc Length to Area (CCH Text 8.2)

Name _____   Due Date _____

We will explore what happens to the ratio of arc length to area on $[0, 1]$ as $a \to \infty$ for four curves that depend on the parameter a. For each of the four functions that follow,

(a) Plot the graph of the function for $a = 1$.

(b) Find the area bounded by the function and the x axis on $[0, 1]$.

(c) With pencil and paper, write down the integral formulas for the arc length on $[0, 1]$ and the area under the curve on $[0, 1]$. Use these to find an integral formula for the limit of the ratio of arc length to area as $a \to \infty$. (<u>Hint</u>: Factor out a from each integral before taking the limit.)

(d) Using your work in Part (c), find the limit as $a \to \infty$ of the ratio of arc length to area on $[0, 1]$.

(e) By looking at the geometry of the graph, can you find a way to predict the limit in Part (d) without doing the calculations?

1. $a(x - x^2)$

2. $a(\frac{1}{2} - |x - \frac{1}{2}|)$

3. $a\sin(\pi x)$

4. a times a semicircle of radius 1 (You figure out the function.)

Solved Problem 8.4: From the Earth to the moon (CCH Text 8.3)

(a) Find the work required to move a rocket ship of mass m from the surface of the Earth to the surface of the moon.

(b) What initial velocity must be given to a cannonball fired from the surface of the Earth if it is to land on the moon?

<u>Solution to (a)</u>: In our solution we are going to take into account the influence of the moon's gravity. The following data will be necessary:

$G =$	Universal gravitational constant	$= 6.67 \times 10^{-11}$
$E =$	Mass of the Earth	$= 5.98 \times 10^{24}$ kilograms
$P =$	Radius of the Earth	$= 6.38 \times 10^{6}$ meters
$L =$	Mass of the moon	$= 7.35 \times 10^{22}$ kilograms
$Q =$	Radius of the moon	$= 1.74 \times 10^{6}$ meters
$D =$	Distance from center of Earth to center of moon	$= 3.84 \times 10^{8}$ meters

In In[1] of Figure 8.5, we have defined all these constants using a list enclosed by set brackets and separated by commas to save space. Note that we use small letters, since E and D are reserved letters in *MATHEMATICA*. (You may find it easier to define each constant on a separate line or to leave them undefined and enter the values at the end of your work.)

If r is the distance from the center of the Earth to the rocket ship, then the force pulling it toward the Earth is $\frac{GEm}{r^2}$, while the force pulling it toward the moon is given by $\frac{GLm}{(D-r)^2}$. Thus, the net force on the rocket is the difference

$$\text{Force} = \frac{GEm}{r^2} - \frac{GLm}{(D-r)^2}$$

To obtain the work, we integrate the force from the surface of the Earth, $r = P$, to the surface of the moon, $r = D - Q$. The resulting integral appears as In[2] of Figure 8.5. We get the work, $5.86702 \times 10^7 m$ joules, in Out[2]. (Recall that a joule is a Newton-meter of work.)

<u>Solution to (b)</u>: The cannonball of mass m must have sufficient kinetic energy, $\frac{1}{2}mv^2$, to pass the point where the gravitational attraction of the moon and the Earth are equal. From that point it will fall to the surface of the moon. To find that point we evaluate **Solve[g*e*m/r^2-g*l*m/(d-r)^2==0,r]**. We see two solutions in Out[3] of Figure 8.5. $4.3188 10^8$ is greater than the distance from the Earth to the moon, so it cannot be correct. We conclude that the

221

```
In[1]:=
    {g,e,p,l,q,d}={6.67 10^(-11),5.98 10^(24),6.38 10^6,
                    7.35 10^22,1.74 10^6,3.84 10^8}
Out[1]=
    {6.67 10^-11, 5.98 10^24, 6.38 10^6, 7.35 10^22, 1.74 10^6,
     3.84 10^8}
In[2]:=
    Integrate[g*e*m/r^2 - g*l*m/(d-r)^2,{r,p,d-q}]
Out[2]=
    5.86702 10^7 m
In[3]:=
    Solve[g*e*m/r^2-g*l*m/(d-r)^2==0,r]
Out[3]=
    {{r -> 3.45677 10^8}, {r -> 4.3188 10^8}}
In[4]:=
    Integrate[g*e*m/r^2 - g*l*m/(d-r)^2,{r,p,3.45677 10^8}]
Out[4]=
    6.12494 10^7 m
In[5]:=
    Solve[(1/2)m*v^2== 6.12494 10^7 m,v]
Out[5]=
    {{v -> -11067.9}, {v -> 11067.9}}
```

Figure 8.5: **Work required to get from the Earth to the moon**

equilibrium point occurs at $r = 3.45677 \times 10^8$ meters. The work required to lift the cannonball to this point is $\int_p^{3.45676 \times 10^8} \frac{GEm}{r^2} - \frac{GLm}{(D-r)^2} \, dr$.

Evaluate **Integrate[g*e*m/r^2-g*l*m/(d-r)^2,{r, p,3.35677 10^8}]** to obtain the work, $6.12494 \times 10^7 m$ joules, as seen in Out[4]. We get the required velocity by setting the kinetic energy equal to this work and solving. We evaluate **Solve[(1/2)m v^2 == 6.12494 10^7 m,v]**. The result is 11067.9 meters per second, as seen in Out[5].

Laboratory Exercise 8.6

From the Earth to the Sun (CCH Text 8.3)

Name _____   Due Date _____

1. Taking into account the gravitational pull of the sun, how much work is required to send a rocket ship of mass m from the surface of the Earth to the surface of the sun? You will need some of the data in Solved Problem 8.4, as well as the following data about the sun:

$$\text{Mass of sun} = 1.97 \times 10^{30} \text{ kilograms}$$
$$\text{Radius of sun} = 6.95 \times 10^{8} \text{ meters}$$
$$\text{Distance from Earth to sun} = 1.49 \times 10^{11} \text{ meters}$$

2. Explain what the sign of the answer means.

3. Find the point between the Earth and the sun where the pull of the Earth's gravity matches that of the sun.

4. Calculate the initial velocity of a cannonball fired from the surface of the Earth if it is to land on the surface of the sun.

5. Calculate the escape velocity for the sun.

Solved Problem 8.5: Rainfall in Anchorage (CCH Text 8.6)

Example 4 in Section 8.6 of the CCH Text uses the following normal probability distribution function with mean 15 and standard deviation 1 to describe the yearly rainfall in Anchorage, Alaska:

$$p(t) := \frac{1}{\sqrt{2\pi}} e^{-\frac{(t-15)^2}{2}}$$

(a) Find the probability that the rainfall in a given year is between 13 and 16 inches.

(b) Plot the graph of the cumulative distribution function.

(c) Find the smallest positive number k so that you can be 98% sure that the rainfall next year in Anchorage will be between $15 - k$ and $15 + k$ inches.

Solution to (a): We evaluate **Clear[p], p[t_]:= (1/Sqrt[2Pi])E ^ (-(t-15) ^2 / 2)** and **Integrate[p[t], { t, 13, 16 }]**. The result is seen in Out[3] of Figure 8.6. The answer is presented in terms of the *error function*, Erf(x). This function is defined as follows.

$$Erf(x) = \frac{2}{\sqrt{\pi}} \int_0^x e^{-t^2} \, dt$$

We evaluate **N[%]** to get the answer, 0.818595, in Out[4].

Solution to (b): The cumulative distribution function is $P(x) = \int_{-\infty}^{x} p(t) \, dt$. We evaluate **Integrate[p[t],{t,-Infinity,x}]** and see that the answer in Out[5] of Figure 8.6 is presented in terms of the Erf function. We defined W[x_] to be this integral and evaluated **Plot[W[x], {x,5,30}]** to view the graph in Out[11].

Solution to (c): The probability that the rainfall is within k inches of 15 is $\int_{15-k}^{15+k} p(t) \, dt$. The integral's evaluation appears in Out[1] of Figure 8.7. We want this probability to be 0.98, so we first try solving for k in Erf($k/\sqrt{2}$) = 0.98, but *MATHEMATICA* cannot solve this exactly. So we evaluate **FindRoot[Erf[k/Sqrt[2]]==0.98,{k,2}]**. (How did we choose 2? Look at Part (c) of Example 4 in the CCH Text.) The solution is $k = 2.32635$. Thus, with 98% certainty we can predict that next year's rainfall in Anchorage will be between 12.67 and 17.32 inches.

You may have noticed that there is a difficulty with the basic assumption that rainfall is normally distributed. This means that we are allowing the possibility that the rainfall next year may be negative! In the output of Figure 8.7 we have calculated the probability allowed for such an event. Do you think the difficulty here is really serious?

```
In[1]:=
   Clear[p]
In[2]:=
   p[t_]:= (1/Sqrt[2Pi])E^(-(t-15)^2/2)
In[3]:=
   Integrate[p[t],{t,13,16}]
Out[3]=
          1
   Erf[-------]
      Sqrt[2]     Erf[Sqrt[2]]
   ------------ + -------------
        2              2
In[4]:=
   N[%]
Out[4]=
   0.818595
In[5]:=
   Integrate[p[t],{t,-Infinity,x}]
Out[5]=
         -15 + x
      Erf[-------]
   1     Sqrt[2]
   - + -----------
   2        2
```

```
In[10]:=
   W[x_]:= 1/2 + Erf[(-15 +x)/Sqrt[2]]/2
In[11]:=
   Plot[W[x],{x,5,30}]
```

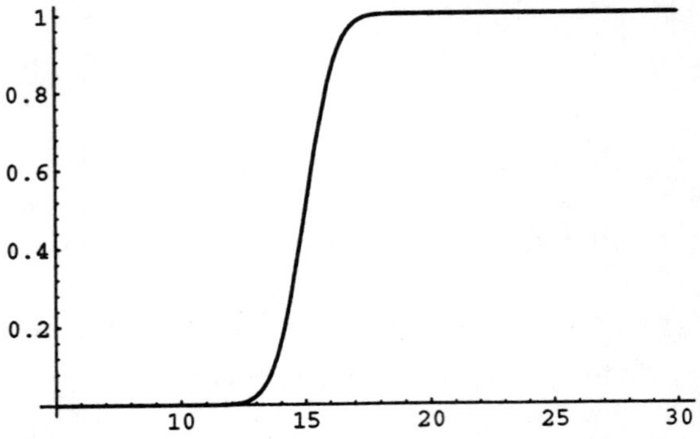

Out[11]=
  -Graphics-

Figure 8.6: **Rainfall in Anchorage**

```
In[12]:=
   Integrate[p[t],{t,15-k,15+k}]
Out[12]=
         k
   Erf[-------]
      Sqrt[2]
In[1]:=
   FindRoot[Erf[k/Sqrt[2]]==0.98,{k,2}]
Out[1]=
   {k -> 2.32635}
```

Figure 8.7: **Predicting rainfall with 98% certainty**

Laboratory Exercise 8.7

SAT Scores (CCH Text 8.6)

Name _____ Due Date _____

According to the book, *For All Practical Purposes*, W.H. Freeman and Co., edited by Lynn Steen, SAT scores are normally distributed with a mean of 500 and a standard deviation of 100.

1. Write the probability distribution function for SAT scores.

2. What percentage of students score between 500 and 550 on the SAT exam?

3. What percentage of students score more than 700 on the SAT exam?

4. Plot the graph of the cumulative distribution function.

5. Find a value of k so that you are 90% certain that a student selected at random will have an SAT score between $500 - k$ and $500 + k$.

6. The assumption of a normal distribution for SAT scores allows the possibility of negative scores on the exam. What percentage of students does this assumption predict will score less than zero? How serious do you think this flaw in the model is?

Chapter 9
Differential Equations

In this chapter we use *MATHEMATICA* as an aid in the qualitative study of differential equations. Although *MATHEMATICA* can solve many differential equations in closed form (that is, exactly) this feature will not be emphasized here.

Solved Problem 9.1: Families of solutions (CCH Text 9.1)

Show that $e^{-x} + ce^{-2x}$ is a family of solutions of $y' + 2y = e^{-x}$. Plot this family of curves.

Solution: The first part of this problem can be solved with pencil and paper, but this exercise will show how to use *MATHEMATICA*. The first step is to write the equation in the form $y' + 2y - e^{-x} = 0$. We evaluate **Clear[y,c,f,eq]** and **eq[y_]:= D[y,x] +2y - E^(-x)**. Next we evaluate **f[x_,c_]:= E^(-x)+c*E^(-2x)**. These appear as inputs 1,2, and 6 of Figure 9.1.

To check the solutions, we substitute them into the differential equation. Evaluate **eq[f[x,c]]** and then **Simplify[%]** to get 0, showing that for each choice of c we get a solution.

To obtain the family of curves, we evaluate **Clear[T]** and **T=Table[f[x,c],{c,-3,3,0.2}]** We get a list of functions in the form of In[6] as c ranges from -3 to 3 in steps of 0.2. (See Out[11].) Evaluate **Plot[Evaluate[T],{x,-2,1}]**, and we see the graphs displayed in Out[12] of Figure 9.1.

In[1]:=
```
Clear[y,c,f,eq]
```
In[2]:=
```
eq[y_] := D[y,x] + 2y - E^(-x)
```
In[6]:=
```
f[x_,c_] := E^(-x) + c*E^(-2x)
```
In[7]:=
```
eq[f[x,c]]
```
Out[7]=
$$\frac{-2\,c}{E^{2\,x}} - \frac{2}{E^{x}} + 2\,(\frac{c}{E^{2\,x}} + E^{-x})$$

In[8]:=
```
Simplify[%]
```
Out[8]=
0

In[10]:=
```
Clear[T]
```
In[11]:=
```
T=Table[f[x,c],{c,-3,3,0.2}]
```
Out[11]=
$$\{\frac{-3}{E^{2\,x}} + E^{-x}, \frac{-2.8}{E^{2\,x}} + E^{-x}, \frac{-2.6}{E^{2\,x}} + E^{-x}, \frac{-2.4}{E^{2\,x}} + E^{-x}, \frac{-2.2}{E^{2\,x}} + E^{-x},$$
$$\frac{-2.}{E^{2\,x}} + E^{-x}, \frac{-1.8}{E^{2\,x}} + E^{-x}, \frac{-1.6}{E^{2\,x}} + E^{-x}, \frac{-1.4}{E^{2\,x}} + E^{-x},$$
$$\frac{-1.2}{E^{2\,x}} + E^{-x}, \frac{-1.}{E^{2\,x}} + E^{-x}, \frac{-0.8}{E^{2\,x}} + E^{-x}, \frac{-0.6}{E^{2\,x}} + E^{-x},$$
$$\frac{-0.4}{E^{2\,x}} + E^{-x}, \frac{-0.2}{E^{2\,x}} + E^{-x}, \frac{3.52366\,10^{-19}}{E^{2\,x}} + E^{-x}, \frac{0.2}{E^{2\,x}} + E^{-x},$$
$$\frac{0.4}{E^{2\,x}} + E^{-x}, \frac{0.6}{E^{2\,x}} + E^{-x}, \frac{0.8}{E^{2\,x}} + E^{-x}, \frac{1.}{E^{2\,x}} + E^{-x},$$
$$\frac{1.2}{E^{2\,x}} + E^{-x}, \frac{1.4}{E^{2\,x}} + E^{-x}, \frac{1.6}{E^{2\,x}} + E^{-x}, \frac{1.8}{E^{2\,x}} + E^{-x},$$
$$\frac{2.}{E^{2\,x}} + E^{-x}, \frac{2.2}{E^{2\,x}} + E^{-x}, \frac{2.4}{E^{2\,x}} + E^{-x}, \frac{2.6}{E^{2\,x}} + E^{-x},$$
$$\frac{2.8}{E^{2\,x}} + E^{-x}, \frac{3.}{E^{2\,x}} + E^{-x}\}$$

In[12]:=
```
Plot[Evaluate[T],{x,-2,1}]
```

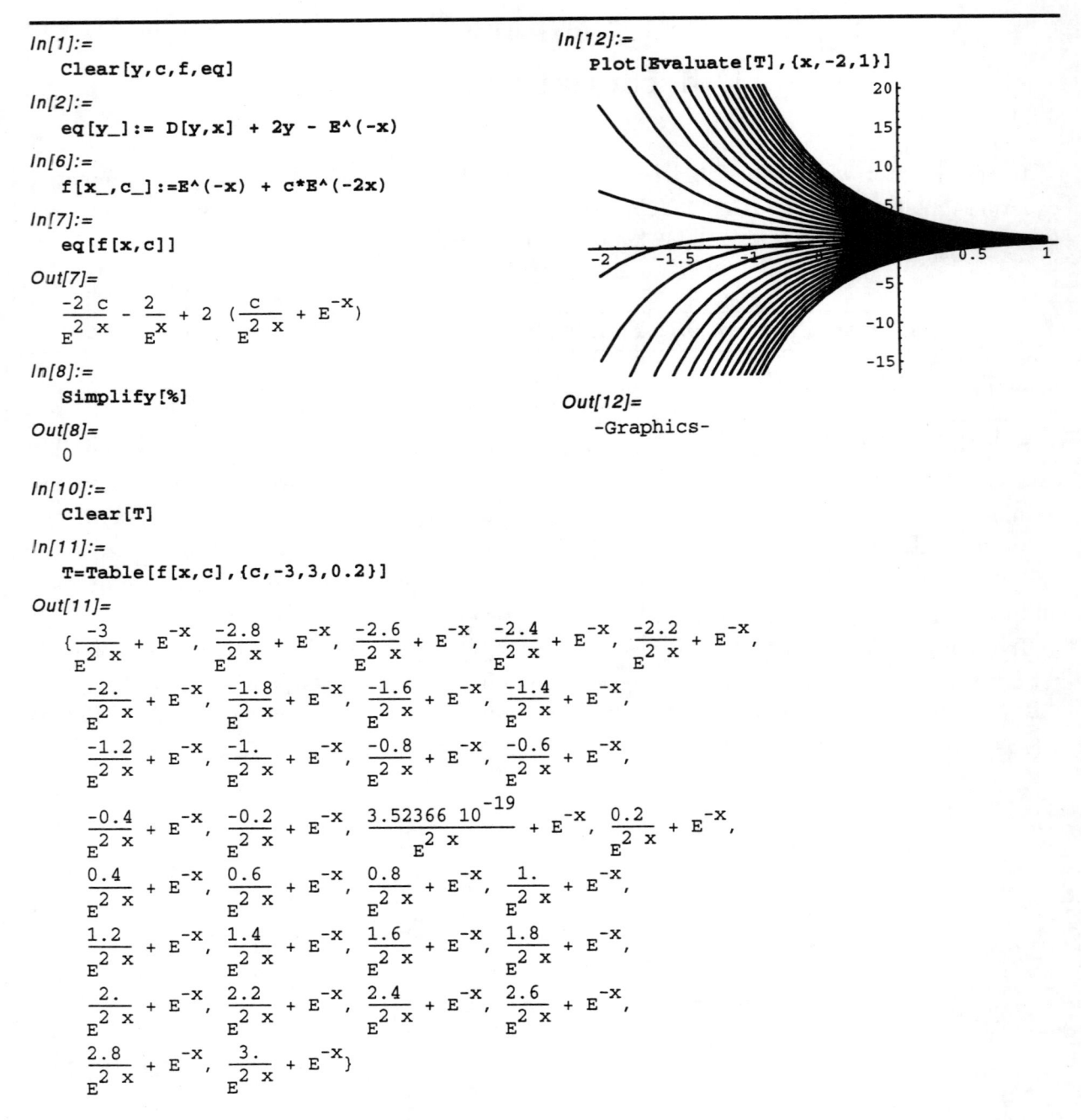

Out[12]=
-Graphics-

Figure 9.1: **Solutions of** $y' + 2y = e^{-x}$

Laboratory Exercise 9.1

Families of Solutions (CCH Text 9.1)

Name _____   Due Date _____

1. Show that $e^{-x}(A \sin x + B \cos x)$ is a family of solutions of $y'' + 2y' + 2 = 0$. (Enter y'' as `D[y,{x,2}]`.)

2. Graph 10 members of each of the two families $Ae^{-x} \sin x$ and $Be^{-x} \cos x$.

3. For what value of k is $e^{-\frac{x}{2}}(A \sin(kx) + B \cos(kx))$ a solution of $y'' + y' + y = 0$? (<u>Hint</u>: Plug $e^{-\frac{x}{2}}(A \sin(kx) + B \cos(kx))$ into the differential equation and see what this tells you about k.)

Solved Problem 9.2: Slope fields (CCH Text 9.2)

Sketch the slope field for the differential equation $2xy + (1+x^2)y' = 0$. Include several solution curves in the picture.

<u>Solution</u>: The function PLOTVECTORFIELD is found in the graphics package 'Plotfiel'. To load this, type

$$\texttt{<<Graphics`PlotField`}$$

(This command works on a Macintosh. For a PC, try `<< graphics`plotfiel`. To find the exact name, open the package folder in the *MATHEMATICA* folder. Open the folder called Graphics or graphics and find the name of the folder closest to PlotField or plotfiel.)

The arguments for the PLOTVECTORFIELD function are similar to the Plot arguments. Assuming your differential equation is in the form $y' = r$, where r is a function of x and y, PlotVectorField[{1,r},{x,a,b,(m)},{y,c,d,(n)}] prepares *MATHEMATICA* to draw directed line segments as x varies from a to b in m steps, and y varies from c to d in n steps. The step values m and n are optional, and you should first try the command without them.

We will make a picture with both x and y varying from -2 to 2. With the package PLOTFIELD.M loaded, we evaluate **PlotVectorField[{1,2x*y/(1+x^2)},{x,-2,2}, {y,-2,2}]**. We see Out[7] of Figure 9.2. This may take a few seconds, depending on your machine.

Once the slope field is produced you can sketch in several solution curves by hand. Your hand–drawn curves should be similar to those displayed in Out[7] of Figure 9.2. (To make the picture in Figure 9.2 we used the exact solution of the equation! If you want to know what it is, consult your *MATHEMATICA* manual on the use of the DSOLVE function.)

In[1]:=
```
<< Graphics`PlotField`
PlotVectorField[{1,-2x*y/(1+x^2)},{x,-2,2},{y,-2,2}]
```

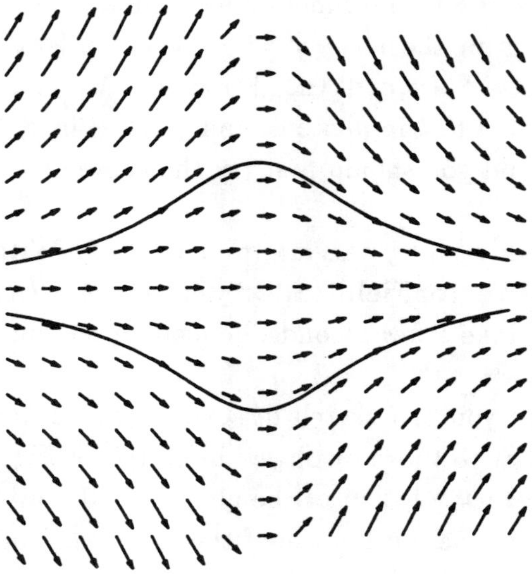

Out[7]=
-Graphics-

Figure 9.2: **Slope field for** $2xy + (1+x^2)y' = 0$

Laboratory Exercise 9.2

Slope Fields (CCH Text 9.2)

Name _____   Due Date _____

For each of the following equations, make a graph of the slope field with *MATHEMATICA* and sketch several solution curves by hand.

1. $(x^2 + y^2)y' = xy$

2. $y' = \sin(xy)$

3. $y' = x^2 - y^2$

Solved Problem 9.3: Euler's method (CCH Text 9.3)

Let $y(x)$ be the solution of $y' = 1 + 2xy$ that passes through the point $(0, -1)$.

(a) Use 5 steps of Euler's method to approximate the graph of $y(x)$ on the interval $x = 0$ to $x = 1$.

(b) Use 20 steps of Euler's method to approximate the graph of $y(x)$ on the interval $x = 0$ to $x = 1$.

(c) Use 20 steps of Euler's method to approximate the value of $y(1)$.

Solution to (a): Load the package MYCALC.m which is given in Appendix II. Appendix II describes how to create this package if you have not already done so. MYCALC.m contains the function euler which implements Euler's method. The syntax for euler is

`euler[f,{x,a,b,stp},{y,start}]`,

where f is a function of x and y. Then euler approximates the function y with derivative, $y' = f(x, y)$ on the interval $[a, b]$ with starting point $(a, start)$. x is the independent variable and y is the dependent variable. stp is the value of Δ, and gives the number of steps for which euler is calculated.

We want to execute 5 steps of Euler's method, beginning at $x = 0$ and ending at $x = 1$. Thus, we choose $stp = 0.2$. Since we want a solution that passes through $(0, -1)$, we set $start = -1$. We must first enter the function f. Evaluate **f[x_,y_]:= 1 + 2x*y**. Then evaluate **euler[1+2x*y,{x,0,1,0.2},{y,-1}]**. See Out[4] of Figure 9.3.

Solution to (b): To get 20 steps from 0 to 1 we choose stp to be 0.05. Now we evaluate **euler[f,{x,0,1,0.05},{y,-1}]**. We get Out[5] of Figure 9.3, a better approximation of the solution.

In[1]:=
    <<MYCALC.m

In[3]:=
    f[x_,y_]:=1+2x*y

In[4]:=
    euler[f,{x,0,1,0.2},{y,-1}]

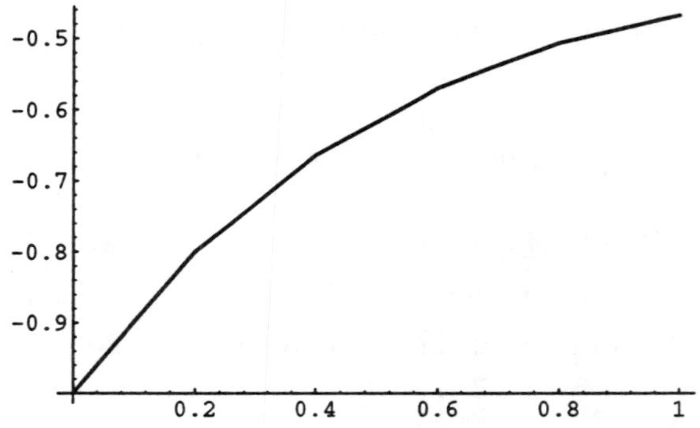

Out[4]=
    -Graphics- && {{0, -1}, {0.2, -0.8}, {0.4, -0.664},
        {0.6, -0.57024}, {0.8, -0.507098}, {1., -0.469369}}

In[5]:=
    euler[f,{x,0,1,0.05},{y,-1}]

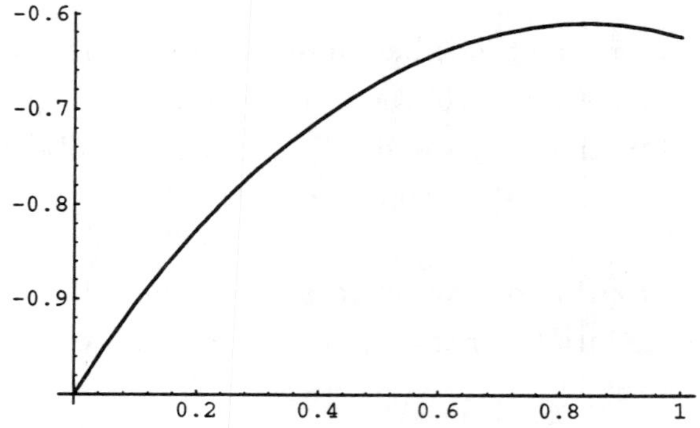

Out[5]=
    -Graphics- && {{0, -1}, {0.05, -0.95}, {0.1, -0.90475},
        {0.15, -0.863797}, {0.2, -0.826754}, {0.25, -0.79329},
        {0.3, -0.763122}, {0.35, -0.736015}, {0.4, -0.711776},
        {0.45, -0.690247}, {0.5, -0.671308}, {0.55, -0.654874},
        {0.6, -0.640892}, {0.65, -0.629345}, {0.7, -0.620253},
        {0.75, -0.61367}, {0.8, -0.609695}, {0.85, -0.608471},
        {0.9, -0.610191}, {0.95, -0.615108}, {1., -0.623544}}

Figure 9.3: **Euler's method for** $y' = 1 + 2xy$

Solution to (c): The last value under the graph is the last point of the graph. So the approximate value of $y(1)$ is -0.623544.

238

Laboratory Exercise 9.3

Estimating Function Values with Euler's Method (CCH Text 9.3)

Name _____   Due Date _____

It is known that the solution curve of the differential equation $y' = \dfrac{y \cos x}{1 + 2y^2}$ that passes through the point $(0, 1)$ satisfies $y(1) = 1.27494$, $y(2) = 1.28718$, and $y(10) = 0.813712$.

1. Execute Euler's method starting at $(0, 1)$ using step size $\Delta x = 0.5$ and $n = 20$. Graph the approximate solution you obtain.

2. What estimates did you get in Part 1 for $y(1)$, $y(2)$, and $y(10)$? Compare these answers with the given values, and discuss how the accuracy of Euler's method varies with the distance from the starting point.

3. Repeat Part 1 using step size $\Delta x = 0.1$ and $n = 100$.

4. What estimates did you get in Part 3 for $y(1)$, $y(2)$, and $y(10)$? Compare these estimates with those you obtained in Part 1 and with the given values. Discuss how the accuracy of Euler's method varies with the step size.

Solved Problem 9.4: A tank of water (CCH Text 9.6)

A water tank initially contains V_0 cubic feet of water. Water runs into the tank at a constant rate of 36 cubic feet per minute, but a valve allows water to flow out of the tank at a rate equal to 20 percent of the volume of water in the tank per minute.

(a) Set up and solve a differential equation that determines the volume of water in the tank as a function of time. Express your solution in terms of t and V_0.

(b) Graph the solutions for $V_0 = 100, 200, \ldots, 1000$.

(c) Find the equilibrium solution. Add the graph of the equilibrium solution to your graph in Part (b) and determine if it is stable or unstable.

<u>Solution to (a)</u> The rate of change of V is the rate the water is coming in minus the rate the water is going out. Therefore,

$$\frac{dV}{dt} = \text{Rate in} - \text{Rate out} = 36 - 0.2V$$

The solution of this equation is $V = 180 - ke^{-0.2t}$ (which you should be able to find by hand). When $t = 0$ the equation is $V_0 = 180 - k$. Thus, $k = 180 - V_0$, and the solution in terms of V_0 is $V = 180 - (180 - V_0)e^{-0.2t}$.

<u>Solution to (b)</u>: We evaluate **Clear[v,v0,T]** and **v[t_,v0_]:=180 - (180 - v0)E^(-0.2t)**. Next we evaluate **T=Table[v[t,v0], { v0, 100, 1000, 100}]** and finally **Plot[Evaluate[T], {t, 0, 10}]**, as displayed in Out[10] of Figure 9.4.

```
In[1]:=
    Clear[v,v0,T]
In[2]:=
    v[t_,v0_]:=180-(180-v0)E^(-0.2t)
In[3]:=
    T=Table[v[t,v0],{v0,100,1000,100}]
Out[3]=
```

$$\{180 - \frac{80}{E^{0.2\,t}},\ 180 + \frac{20}{E^{0.2\,t}},\ 180 + \frac{120}{E^{0.2\,t}},\ 180 + \frac{220}{E^{0.2\,t}},$$
$$180 + \frac{320}{E^{0.2\,t}},\ 180 + \frac{420}{E^{0.2\,t}},\ 180 + \frac{520}{E^{0.2\,t}},\ 180 + \frac{620}{E^{0.2\,t}},$$
$$180 + \frac{720}{E^{0.2\,t}},\ 180 + \frac{820}{E^{0.2\,t}}\}$$

```
In[4]:=
    Plot[Evaluate[T],{t,0,10}]
```

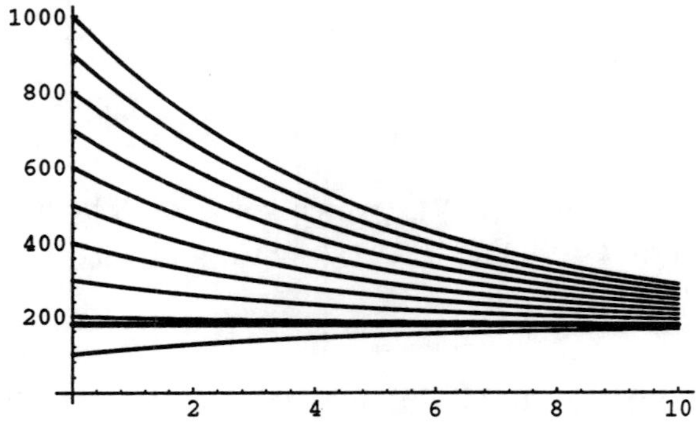

```
Out[10]=
    -Graphics-
```

Figure 9.4: **Volume of water in a tank**

Solution to (c): The equilibrium solution occurs when $\frac{dV}{dt} = 0$, that is, when $V = 180$. (Note that we obtained this without resorting to the solution of the differential equation.) The horizontal line $V = 180$ (the dark line) has been added to Figure 9.4. From the graphs we see that solutions both above and below bend toward the equilibrium solution. We conclude that the equilibrium solution is stable. We can also verify this analytically by noting that for any V_0, $\lim_{t\to\infty} 180 - (108 - V_0)e^{-0.2t} = 180$.

Laboratory Exercise 9.4

A Leaky Balloon (CCH Text 9.5)

Name _____   Due Date _____

A leaky balloon initially contains V_0 cubic inches of air. Additional air is being blown into the balloon at a constant rate of 24 cubic inches per minute. The balloon leaks air through a small hole at a rate equal to 3 percent of its volume per minute at any given time.

1. Set up and solve a differential equation that gives the volume of air in the balloon as a function of time t. Express your answer in terms of t and V_0

2. Find the equilibrium solution, and explain what it means in practical terms.

3. Graph the equilibrium solution as well as several other nearby solution curves.

4. Is the equilibrium solution stable? Justify your answer both analytically and graphically.

5. What is the diameter of the balloon when the equilibrium solution is attained?

Laboratory Exercise 9.5

Baking Potatoes (CCH Text 9.5)

Name _____   Due Date _____

A chef realizes at 6:30 that he has forgotten to preheat the oven for baked potatoes to be served with dinner at 7:30. He turns on the oven and puts in the potatoes. The oven temperature t minutes after it is turned on is $O(t) = 400 - 325e^{-0.7t}$ degrees Fahrenheit. The potatoes heat according to Newton's law: The rate of change in the temperature, $P(t)$, of the potatoes is $\frac{dP}{dt} = 0.03(O(t) - P(t))$. The potatoes will be done when they reach a temperature of 270 degrees.

1. Set up a differential equation whose solution gives the temperature of the potatoes at time t. (Do not try to solve it.)

2. What is the temperature of the oven when $t = 0$? (You may assume this is also the initial temperature of the potatoes.)

3. Use Euler's method to approximate and graph a solution of the equation. We recommend a step size of 0.2. Get your starting point from Part 2 and choose n large enough to find out when the potatoes are done. (If they aren't done in an hour, the chef will probably be fired.)

4. According to your solution, when will the potatoes be done?

5. What time would the potatoes have been done if the chef had remembered to preheat the oven to 400 degrees? (You should be able to set up and solve a differential equation to answer this.)

6. (**Optional**): Perform an experiment at home to determine if the times predicted in this exercise are reasonable.

Solved Problem 9.5: Terminal velocity (CCH Text 9.6)

The velocity (in feet per second) of a falling object subject to air resistance is governed by the differential equation $\frac{dv}{dt} = 32 - 0.4v^{\frac{3}{2}}$.

(a) What is the terminal velocity of the object?

(b) Use Euler's method to produce a graph of velocity as a function of time if the initial velocity is zero.

(c) How long does it take for the object to reach 99% of its terminal velocity?

<u>Solution to (a)</u>: Terminal velocity occurs when the rate of change in velocity is zero, that is, when $32 - 0.4v^{\frac{3}{2}} = 0$. Solving this equation gives $v = 18.5663$.

<u>Solution to (b)</u>: We know from experience that the object will approach its terminal velocity quickly, so we use Euler's method on the interval $t = 0$ to $t = 3$. Load the package MYCALC.m. If you have not created it, see Appendix II. Now we **Clear[v,dv]** and define **dv[t_,v_]:= 32 - 0.4 v^(3/2)**. Next evaluate **euler[dv,{t,0,3,0.01},{v,0}]** as below in In[8] of Figure 9.5. (We have suppressed the numerical output.) From the numerical output and the graph we see that the terminal velocity is approximately 18.5565 feet per second.

<u>Solution to (c)</u>: 99% of the terminal velocity found in Part (a) is 18.3709. Examining the graph (which is omitted in Figure 9.5), we see that 99% of the terminal velocity is reached at approximately 1.87 seconds.

In[6]:=
```
<< MYCALC.m
```
In[7]:=
```
Clear[v,dv]
```
In[8]:=
```
dv[t_,v_] := 32-0.4v^(3/2)
```
In[14]:=
```
euler[dv,{x,0,3,0.01},{v,0}]
```

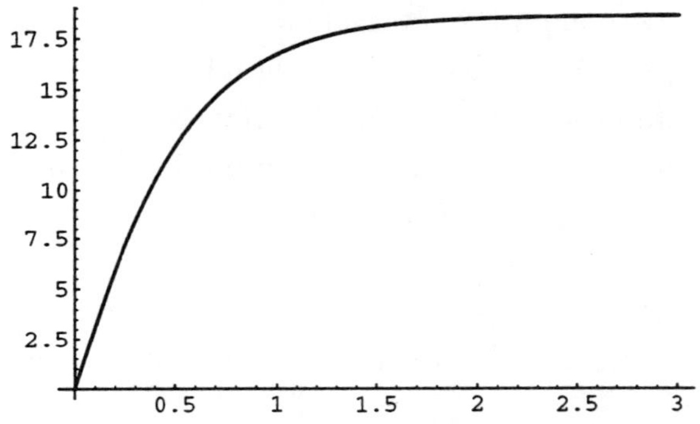

In[15]:=
```
.99*18.5565
```
Out[15]=
```
18.3709
```

Figure 9.5: **Estimating time to terminal velocity**

Laboratory Exercise 9.6

Drag and Terminal Velocity (CCH Text 9.6)

Name _____   Due Date _____

This is a continuation of Solved Problem 9.5.

1. The coefficient 0.4 of $v^{3/2}$ in Solved Problem 9.5 represents *drag*. Use Euler's method to produce the graph of velocity against time if the drag is doubled and if it is tripled. Add both to the graph from Solved Problem 9.5.

2. How long does it take to reach 99% of terminal velocity if the drag is doubled? How long if the drag is tripled?

3. A feather and a cannonball are dropped at the same time. Which reaches 99% of its terminal velocity first? Explain.

Solved Problem 9.6: Population growth with a threshold (CCH Text 9.7)

A population, $P(t)$ at time t, is governed by the following differential equation:

$$\frac{dP}{dt} = -P(1 - \frac{P}{500})(1 - \frac{P}{2000})$$

(a) Find the equilibrium solutions. Plot $\frac{dP}{dt}$ as a function of P and use the graph to classify each equilibrium solution as stable or unstable.

(b) Use Euler's method to obtain approximate solutions for initial populations $P(0) = 200, 400, 600, 1000, 2500,$ and 3000. Graph each of these solutions together with the equilibrium solutions found in Part (a).

(c) Explain in practical terms what these graphs say about the population in question.

<u>Solution to (a):</u> Equilibrium solutions occur where $\frac{dP}{dt} = 0$, that is, when $-P(1 - \frac{P}{500})(1 - \frac{P}{2000}) = 0$, or when P is 0, 500 or 2000.

To get the graph and to use Euler's method in part (b), we **Clear[P,dP]** and define **dP[P_]:= -P(1-P/500)(1-P/2000)**.

To get the graph of $\frac{dP}{dt}$ as a function of P we evaluate **Plot[dP[P],{P,-100,2300}]**. The result is in Out[9] of Figure 9.6. (It is worth repeating the warning in the CCH Text that this is *not* the graph of P as a function of t; rather, it is the graph of the rate of change in population as a function of population.)

From the graph in Out[9] of Figure 9.6, we see that when $P < 500$, $\frac{dP}{dt}$ is negative, so that P is a decreasing function of t. Thus, if $P < 500$, the population decreases toward 0, and we conclude that $P = 0$ is a stable equilibrium solution.

On the interval $500 < P < 2000$, $\frac{dP}{dt}$ is positive, and thus P is increasing on this interval. But P is decreasing for $P > 2000$. From this analysis, populations between 500 and 2000 will increase toward 2000, and populations over 2000 will decrease toward 2000. Thus $P = 500$ is an unstable equilibrium solution, and $P = 2000$ is a stable equilibrium solution.

<u>Solution to (b):</u> We load MYCALC.m by evaluating **<< CALC.m**, assuming you created it. (To create MYCALC.m, see Appendix II.) We will use EULERPIC, which just gives the graphical representation of Euler's method. To prepare to use EULERPIC, we need a function of t and P. Therefore, evaluate **Clear[dp]** and **dp[t_,P_]:= dP[P]**. See In[15] in Figure 9.7. We want to create a number of graphs, so we create a list of the plots. First we create a function that creates the plots for different starting values: **el[x_]:= eulerpic[dp,{t,0,5,0.05},{P,x}]**. Note that we have use the interval $[0, 5]$ with $\Delta t = 0.05$. The variable x is the starting value for P. See

```
Clear[P,dP]
In[12]:=
    dP[P_]:=-P(1-P/500)(1-P/2000)
    Plot[dP[P],{P,-100,2300}]
```

[Plot showing dP/dt vs P with axes marked at 500, 1000, 1500, 2000 on horizontal axis and 500, -500, -1000 on vertical axis]

-Graphics-

Figure 9.6: $\dfrac{dP}{dt}$ as a function of P

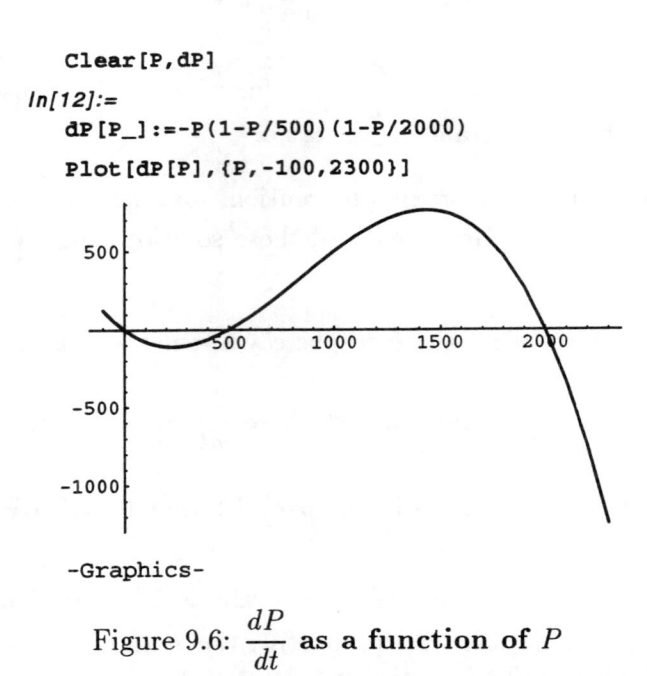

In[15] in Figure 9.7. Next evaluate **T={el[200],el[400],el[600],el[1000],el[2500],el[3000]}**. We have suppressed the output from these. Finally, to get the desired graph, evaluate **Show[T]**, which produces the graphs displayed in Out[18] of Figure 9.7.

Notice that the solution curves above and below $P = 2000$ tend toward that solution, verifying that it is stable, while curves near $P = 500$ move away, indicating that this solution is unstable. This concurs with our earlier classification of the equilibrium solutions.

```
In[13]:=
    Clear[dp]
    dp[t_,P_] := dP[P]
In[14]:=
    << MYCALC.m
In[15]:=
    el[x_]:=eulerpic[dp,{t,0,5,0.05},{P,x}]
In[17]:=
    T={el[200],el[400],el[600],el[1000],el[2500],el[3000]}
In[18]:=
    Show[T]
```

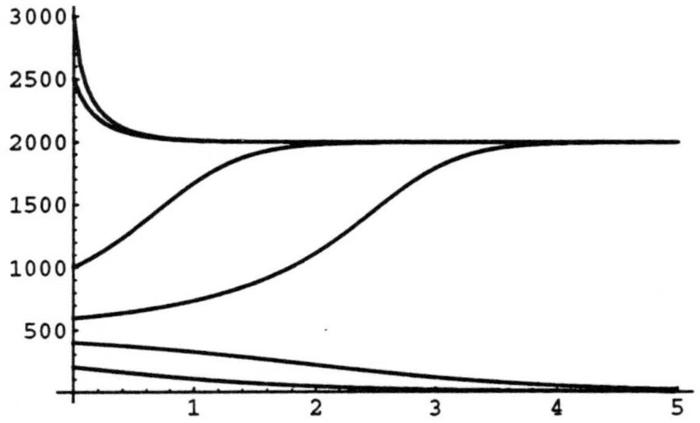

Out[18]=
  -Graphics-

Figure 9.7: **Equilibrium and other solutions**

Solution to (c): This is an example of a population model with a *threshold*. If the population ever drops below 500 individuals, it will eventually die out (perhaps because there is an insufficient number of mating pairs to sustain the population).

If the population is above the threshold of 500 but below 2000, it can be expected to increase toward 2000, but populations above 2000 quickly decrease to that value. This indicates that a population level of 2000 may be the largest that the environment can support, and that if we start with a population of more than 500 individuals we can expect to eventually see the population stabilize near 2000.

Laboratory Exercise 9.7

Comparing Population Models (CCH Text 9.7)

Name _____   Due Date _____

The Gompertz equation, $\frac{dP}{dt} = 0.7 P \ln\left(\frac{2000}{P}\right)$, is used for population modeling.

1. Graph $\frac{dP}{dt}$ as a function of P. Find the equilibrium points and determine which are stable and which are unstable.

2. Use Euler's method to approximate 10 solutions of the Gompertz equation. Graph these solutions along with all equilibrium solutions.

3. Compare the Gompertz model with the logistic model discussed in Solved Problem 9.5. Your discussion should include equilibrium solutions, thresholds, carrying capacity, and rates of population growth or decline.

Solved Problem 9.7: Spider mites and ladybugs (CCH Text 9.8)

Spider mites infest a gardener's plot of marigolds. Ladybugs eat spider mites. The number of spider mites, S, and ladybugs, L, are governed by the following system of differential equations.

$$\frac{dS}{dt} = 2S - SL$$
$$\frac{dL}{dt} = -6L + 3SL$$

(a) Find the equilibrium points.

(b) Plot the slope field for the phase plane near the equilibrium points, and include several trajectories in your picture.

(c) Discuss how the two populations can be expected to vary with time.

Solution to (a): The equilibrium points occur where both $2S - SL = 0$ and $-6L + 3SL = 0$. This system of equations is easily solved to obtain the equilibrium points $(0,0)$ and $(2,2)$.

Solution to (b): To obtain a slope field for the phase plane, we must consider

$$\frac{dS}{dL} = \frac{2S - SL}{-6L + 3SL}$$

We load the PlotVectorField function with **<< Graphics'PlotField'**. (The package may be different on a PC.) Now evaluate **PlotVectorField[{ 1, (2s-2l)/(-6l+3s*l)}, { l,1,3,0.13}, { s, 1, 3, 0.23}]**. The slope field appears in Out[7] of Figure 9.8. We used 0.13 and 0.23 to avoid the case where $-6l + 3sl = 0$. A look at the point $L = 2$, $S = 1$, shows that $\frac{dL}{dt}$ is negative. Thus the trajectories move in a counterclockwise direction.

257

```
In[1]:=
    <<Graphics`PlotField`
In[6]:=
    PlotVectorField[{1,(2s-s*l)/(-6l+ 3s*l)},{l,1,3,0.13},{s,1,3,.23}]
```

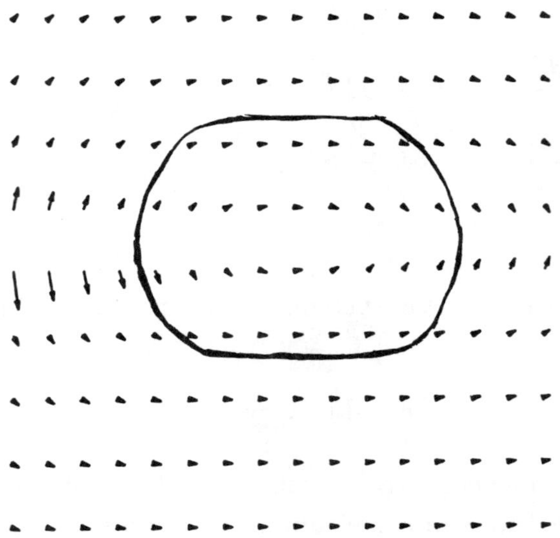

Out[6]=
    -Graphics-

Figure 9.8: **Spider mites and ladybugs**

Solution to (c): The trajectories appear to be ellipses centered around the equilibrium point $(2,2)$. As with the example in Section 9.8 of the CCH Text involving robins and worms, we expect the populations to vary cyclically. If we are at the bottom of one of the ellipses, there aren't enough spider mites (the y axis) to feed the ladybugs, so ladybugs die and the spider mite population increases. Moving counterclockwise, toward the left-hand end of the ellipse, we see that there are enough spider mites to reproduce rapidly and enough to support the diminished ladybug population. Thus, both populations increase. At the top of the ellipse, there are so many spider mites that the ladybug population can indulge itself at the expense of the spider mites. Finally at the right-hand end of the ellipse, the spider mite population has dwindled to the point where the large population of ladybugs cannot get enough food. Both populations decrease until we return once more to the bottom.

Laboratory Exercise 9.8

Foxes and Hares (CCH Text 9.8)

Name _____   Due Date _____

Foxes prey on hares. The populations of foxes, $F(t)$, and hares, $H(t)$, are governed by the following system of equations:

$$\frac{dF}{dt} = 3F - 2FH$$
$$\frac{dH}{dt} = -5H + 2FH$$

1. Find the equilibrium points.

2. Plot the slope field for the phase plane near the equilibrium points, and add several trajectories to your picture.

3. Discuss how the two populations can be expected to vary with time.

Solved Problem 9.8: Springs and second-order equations (CCH Text 9.10)

The position of a moving spring at time t is given by $A\sin(\omega t + \phi)$. Investigate what happens as the constants A, ω, and ϕ vary. How do they relate to the physical characteristics of the spring and its motion?

<u>Solution</u>: We set $\omega = \phi = 1$ and examine graphs as A varies. To get the picture in Out[3] of Figure 9.9, evaluate **Plot[Evaluate [Table[a*Sin[1+t], { a, 0, 2, 0.5}]], { t, -2Pi, 2Pi }]**. Be careful of your brackets.

Notice that as A increases, the amplitude of the curve increases, but nothing else changes. Thus, larger values of A correspond to a larger maximum displacement of the spring.

To examine changes in the phase shift, ϕ, we put $A = \omega = 1$ and vary ϕ. To get the picture in Out[6] of Figure 9.9 we evaluate **Plot[Evaluate [Table[Sin[p+t], { p, 0, 2, 0.5}]], { t, -2Pi, 2Pi }]**.

Notice that changing the phase shift changes the time when maximum displacement of the spring occurs.

Finally, we evaluate **Plot[Evaluate [Table[Sin[1+w*t], { w, 0, 2, 0.5}]], { t, -2Pi, 2Pi }]**. To examine the effect of ω. Notice from Out[9] of Figure 9.9 that as ω increases the period of oscillation decreases. Thus, the parameter ω determines the rate of oscillation of the spring; larger values of ω correspond to a stronger spring.

In[4]:=
```
Plot[Evaluate[Table[a*Sin[1+t],{a,0,2,0.5}]],{t,-2Pi,2Pi}]
```

Out[4]=
-Graphics-

In[6]:=
```
Plot[Evaluate[Table[Sin[p+t],{p,0,2,0.5}]],{t,-2Pi,2Pi}]
```

Out[6]=
-Graphics-

In[9]:=
```
Plot[Evaluate[Table[Sin[1+w*t],{w,0,2,0.5}]],{t,-2Pi,2Pi}]
```

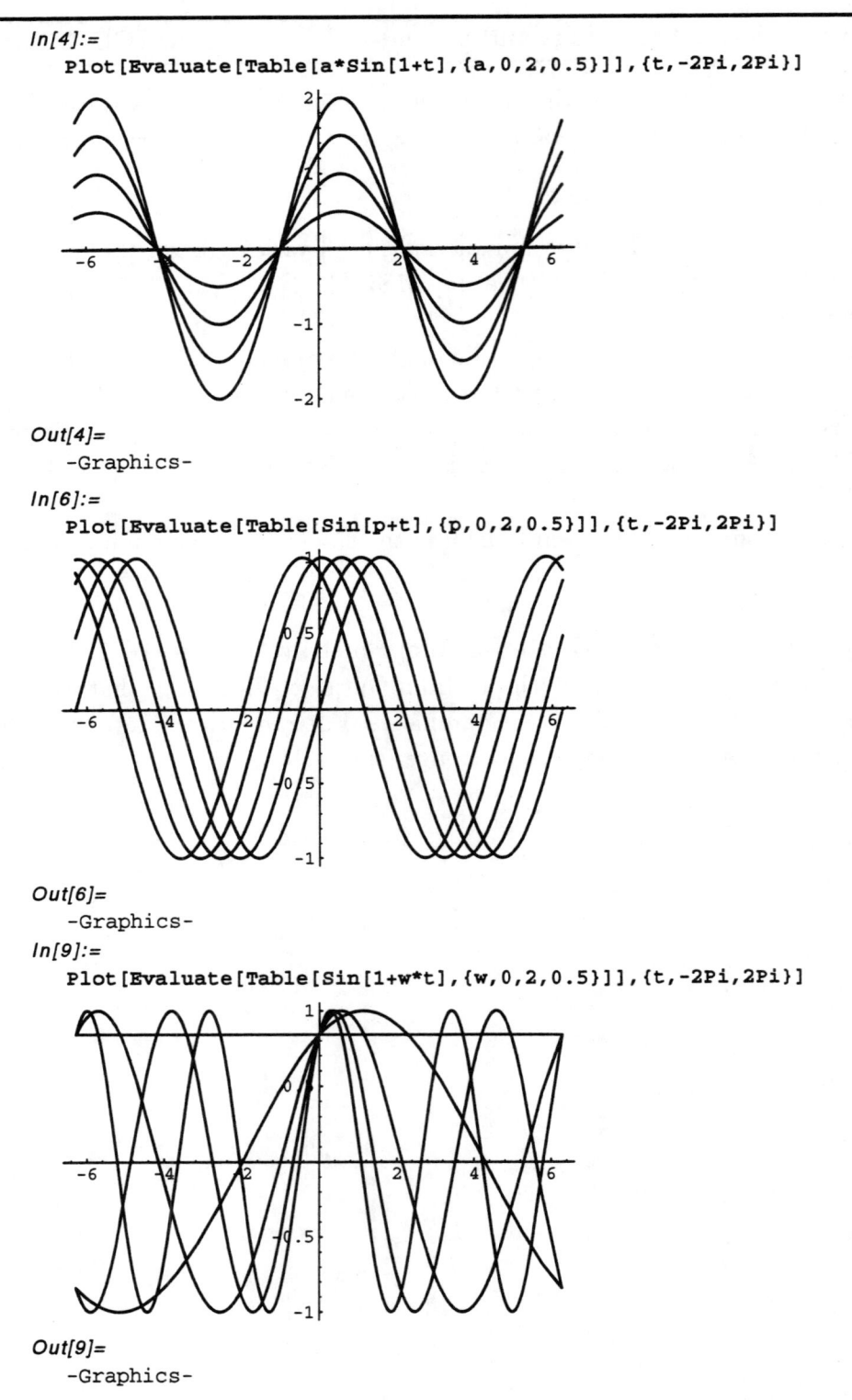

Out[9]=
-Graphics-

Figure 9.9: **Varying the amplitude, period, and rate of oscillation**

Laboratory Exercise 9.9

A Damped Spring (CCH Text 9.11)

Name _____   Due Date _____

In this laboratory, we will explore the movement of a spring in the underdamped, critically damped, and overdamped cases.

1. In the underdamped case, the movement of the spring is given by $e^{-at}(C_1 \cos Bt + C_2 \sin Bt)$. We will examine the special case, $C_1 = C_2 = 1$.

    (a) What is the initial displacement of the spring?
    (b) What is the initial velocity of the spring?
    (c) Produce graphs to show the effects of varying a and B.
    (d) Discuss the physical interpretation of a and B.

2. In the critically damped case, the movement of the spring is given by $(C_1 + C_2 t)e^{-at}$. Again, we specialize to the case $C_1 = C_2 = 1$.

    (a) What is the initial displacement of the spring?
    (b) What is the initial velocity of the spring?
    (c) Produce graphs to show the effects of varying a.
    (d) What is the physical interpretation of a?

3. In the overdamped case, the movement of the spring is given by $C_1 e^{At} + C_2 e^{Bt}$. We take $C_1 = C_2 = 1$.

    (a) What is the initial displacement of the spring?
    (b) What is the initial velocity of the spring?
    (c) Produce graphs to show the effect of varying A and B.
    (d) What are the physical interpretations of A and B?

Chapter 10
Approximations

In this chapter we examine methods of approximating functions by polynomials and extend this idea to Taylor series. We also look at Fourier series, which are expansions in terms of periodic functions.

Solved Problem 10.1: Approximating $\cos x$ (CCH Text 10.1)

(a) Find the tenth–degree Taylor polynomial, $P(x)$, for $\cos x$ about $x = 0$ and plot the two graphs as in Figure 10.1.

(b) Compare the values of $\cos x$ and $P(x)$ for $x = 1, 3$, and 5. Discuss how the size of x affects the accuracy of the approximation.

In[2]:=
```
Series[Cos[x],{x,0,10}]
```
Out[2]=
$$1 - \frac{x^2}{2} + \frac{x^4}{24} - \frac{x^6}{720} + \frac{x^8}{40320} - \frac{x^{10}}{3628800} + O[x]^{11}$$

In[3]:=
```
Normal[Series[Cos[x],{x,0,10}]]
```
Out[3]=
$$1 - \frac{x^2}{2} + \frac{x^4}{24} - \frac{x^6}{720} + \frac{x^8}{40320} - \frac{x^{10}}{3628800}$$

In[4]:=
```
Plot[{Cos[x],%},{x,-2Pi,2Pi}]
```

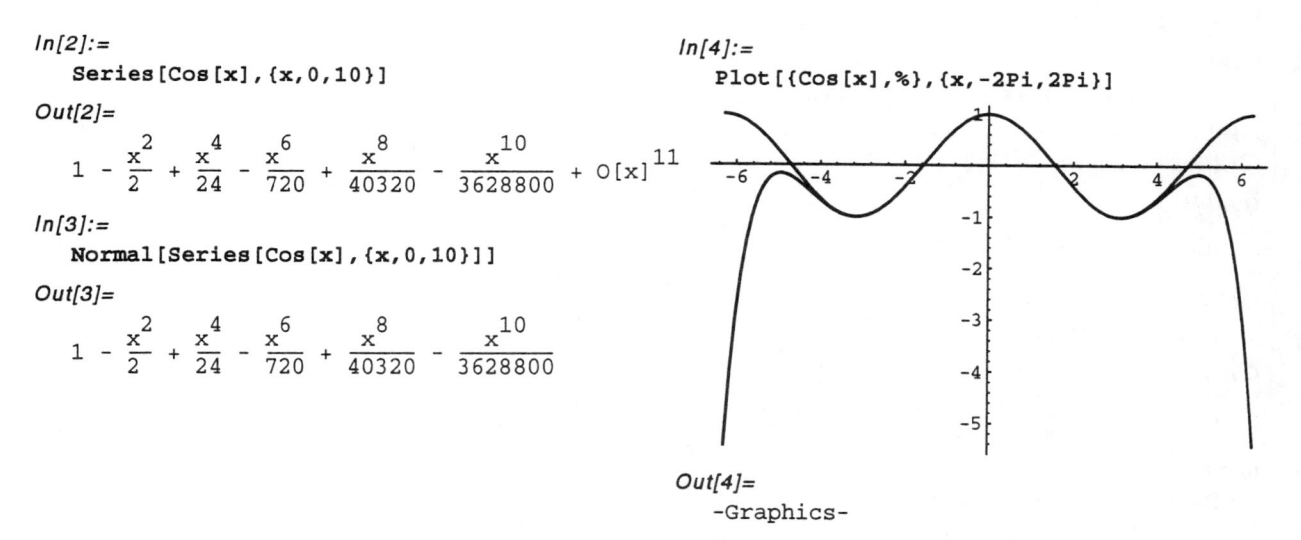

Out[4]=
-Graphics-

Figure 10.1: $\cos x$ and its tenth–order Taylor polynomial

Solution to (a): We use *MATHEMATICA*'s Series function. The syntax of this function is Series[expr,{x,x_0,n}], where *expr* is a function of x, x_0 is the center, and n is the degree of the Taylor polynomial. See the *MATHEMATICA* manual for more details. We evaluate

Series[Cos[x],{x,0,10}] to get Out[2] of Figure 10.1. The inclusion of $O[x]^{11}$ shows that there is an error term. The Taylor polynomial is the output omitting the error term. *MATHEMATICA* will not graph the output of the Series command. To remove the error we use the Normal function. Evaluate **Normal[Series[Cos[x],{x,0,10,}]** and you see that the error is no longer present. See Output[3] of Figure 10.1. Now we can show both graphs by evaluating **Plot[{Cos[x],% },{x,-2Pi,2Pi}]** as seen in Out[4] of Figure 10.1. We leave it to the reader to determine which one is which.

Solution to (b): One commonly used tool to judge the quality of approximations is the *relative error*. Define p[x] as the Taylor polynomial. See In[5] of Figure 10.2. In our case, it is $\frac{\cos x - p[x]}{\cos x}$. This is useful because if we take the absolute value and multiply by 100, we get the error as a percentage of the size of the function. So let's **Clear[er,p]**, and define p[x] as the Taylor polynomial. See In[6] of Figure 10.2. Evaluate **er[x_]:= N[(Cos[x]-p[x])/Cos[x]]**, as seen in In[8] of Figure 10.2.

In[6]:=
  p[x_]=Normal[Series[Cos[x],{x,0,10}]]

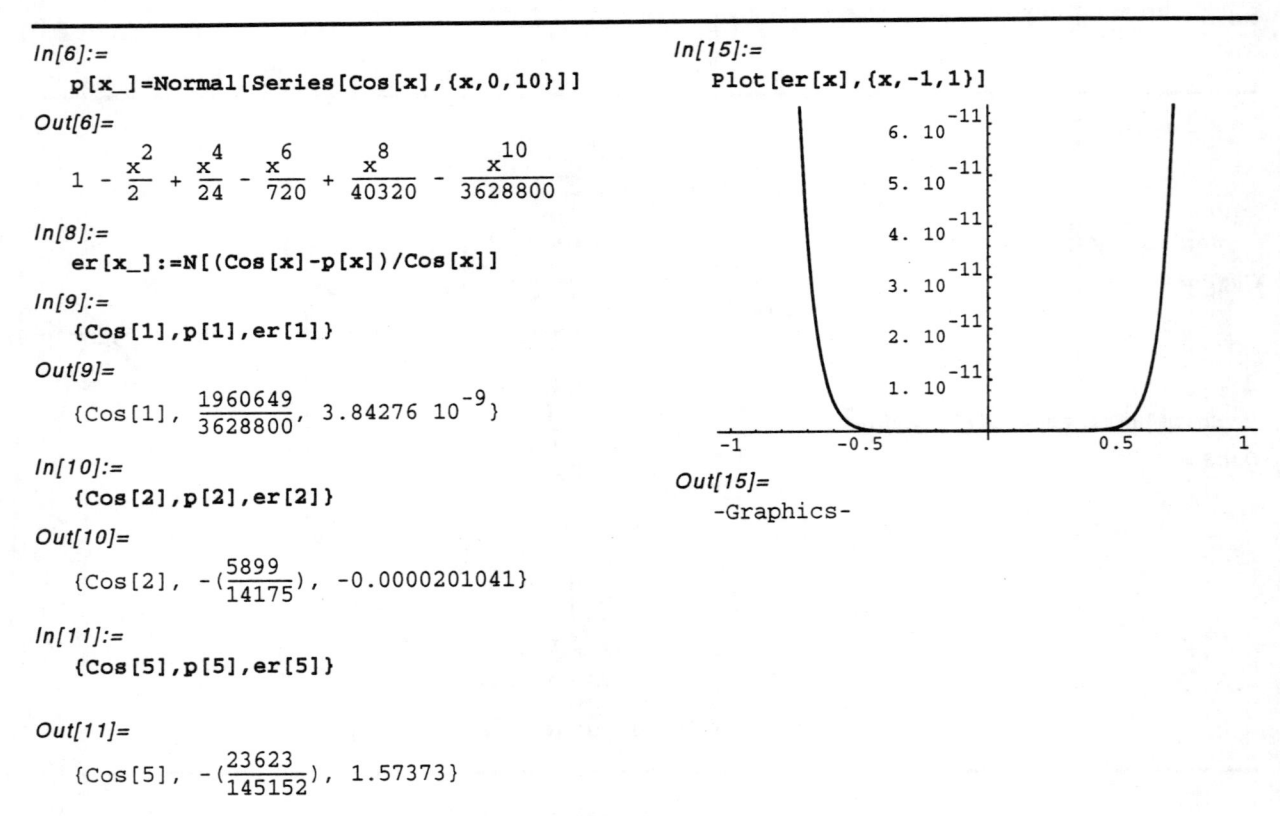

Out[6]=
  $1 - \frac{x^2}{2} + \frac{x^4}{24} - \frac{x^6}{720} + \frac{x^8}{40320} - \frac{x^{10}}{3628800}$

In[8]:=
  er[x_]:=N[(Cos[x]-p[x])/Cos[x]]

In[9]:=
  {Cos[1],p[1],er[1]}

Out[9]=
  {Cos[1], $\frac{1960649}{3628800}$, 3.84276 10^{-9}}

In[10]:=
  {Cos[2],p[2],er[2]}

Out[10]=
  {Cos[2], $-(\frac{5899}{14175})$, -0.0000201041}

In[11]:=
  {Cos[5],p[5],er[5]}

Out[11]=
  {Cos[5], $-(\frac{23623}{145152})$, 1.57373}

In[15]:=
  Plot[er[x],{x,-1,1}]

Out[15]=
  -Graphics-

Figure 10.2: **Error in the Taylor polynomial for the cosine**

We can see the values $\cos(1)$, $p(1)$, and the relative error if we evaluate {**Cos[1],p[1],er[1]**}. In Out[9] we see that $p(1)$ is only off by $3.8427610^{-9}\%$. From Out[10] and Out[11], $p(2)$ misses by 0.0020104%, but $p(5)$ is off by more than 150%. It appears that $p(x)$ provides good approximations for $\cos x$ near the origin, but gives poor approximations for larger values of x. In Out[15] of Figure 10.2 we have plotted the relative error.

Laboratory Exercise 10.1

Taylor Polynomials and the Cosine Function:
The Expansion Point (CCH Text 10.1)

Name _____   Due Date _____

In this exercise you will explore two strategies for approximating $\cos x$ for large values of x.

1. Let $P(x)$ be the eighteenth–degree Taylor polynomial for the cosine function expanded about $x = 0$. Plot the graphs of $\cos x$ and $P(x)$ on the same screen.

2. Plot the graph of the relative error function, and discuss the accuracy of using $P(x)$ as an approximation for $\cos x$ on the interval $[-\pi,\ \pi]$.

3. Since the cosine function has period 2π, $\cos(10) = \cos(10 - 4\pi)$. Show how to use this fact to obtain an accurate approximation of $\cos(10)$ using the Taylor polynomial from Part 1. Calculate the relative error of the approximation.

4. Use the Taylor polynomial from Part 1 to obtain an approximation for $\cos(2438762)$, and report the relative error. (<u>Hint</u>: Solve $2k\pi = 2438762$, choose an integer near the solution, and then follow the idea in Part 3.)

5. An alternative strategy for approximating $\cos(10)$ is to expand the cosine function about a point that is closer to 10.

   (a) Find the sixth–degree Taylor polynomial for $\cos x$ using expansion point 3π.

   (b) Use the Taylor polynomial you found in Part (a) to approximate $\cos(10)$ and report the relative error.

Laboratory Exercise 10.2

Taylor Polynomials and the Cosine Function: The Degree (CCH Text 10.2)

Name _____   Due Date _____

1. On the same screen plot $\cos x$ and its Taylor polynomials about $x = 0$ of degrees 0 through 10. You can make all the Taylor polynomials at once by setting $f(x) = \cos x$, $n = 10$, and $x_0 = 0$; then create a list of functions via **T = Table[NormalSeries[Cos[x],{x,0,i}], {i,0,10 }]**. Then use Plot[Evaluate[T],...].

2. Discuss the relationship between the cosine function and its Taylor polynomial of degree n as n increases.

3. Find (by trial and error) the Taylor polynomial about $x = 0$ of smallest degree for the cosine function that approximates $\cos 10$ to 3 digits of accuracy.

4. On the same screen, plot $\ln x$ and its Taylor polynomials about $x = 1$ of degrees 0 through 10.

5. Discuss the relationship between the function $\ln x$ and its Taylor polynomial of degree n as n increases. Compare the picture here with that in Part 1, and explain any similarities *and differences* you observe between the behavior of the Taylor polynomials in the two cases.

Solved Problem 10.2: Interval of convergence (CCH Text 10.2)

Plot $\dfrac{1}{x^2+7}$ and several of its Taylor polynomials about $x = 0$. Use this to estimate the interval of convergence of its Taylor series.

Solution: We evaluate **Plot[1/(x^2+7),{x,-10,10}]** to get a better picture. To list the first 10 Taylor polynomials at once, we clear f and set $f(x) = 1/(x^2+7)$. Then we clear T and evaluate **T=Table[Normal[Series[f[x],{ x, 0, k}]],{k, 0, 10}]**, which creates all the Taylor polynomials at once. Now evaluate **Plot[Evaluate[T],{x,-10,10}]** to get the graph in Out[4] of Figure 10.3.

In[1]:=
  f[x_]:=1/(x^2+7)

In[2]:=
  Plot[f[x],{x,-10,10}]

Out[2]=
  -Graphics-

In[3]:=
  T=Table[Normal[Series[f[x],{x,0,k}]],{k,0,10}]

Out[3]=

$\{\dfrac{1}{7}, \dfrac{1}{7}, \dfrac{1}{7} - \dfrac{x^2}{49}, \dfrac{1}{7} - \dfrac{x^2}{49}, \dfrac{1}{7} - \dfrac{x^2}{49} + \dfrac{x^4}{343}, \dfrac{1}{7} - \dfrac{x^2}{49} + \dfrac{x^4}{343},$

$\dfrac{1}{7} - \dfrac{x^2}{49} + \dfrac{x^4}{343} - \dfrac{x^6}{2401}, \dfrac{1}{7} - \dfrac{x^2}{49} + \dfrac{x^4}{343} - \dfrac{x^6}{2401},$

$\dfrac{1}{7} - \dfrac{x^2}{49} + \dfrac{x^4}{343} - \dfrac{x^6}{2401} + \dfrac{x^8}{16807}, \dfrac{1}{7} - \dfrac{x^2}{49} + \dfrac{x^4}{343} - \dfrac{x^6}{2401} + \dfrac{x^8}{16807},$

$\dfrac{1}{7} - \dfrac{x^2}{49} + \dfrac{x^4}{343} - \dfrac{x^6}{2401} + \dfrac{x^8}{16807} - \dfrac{x^{10}}{117649}\}$

In[4]:=
  Plot[Evaluate[T],{x,-10,10}]

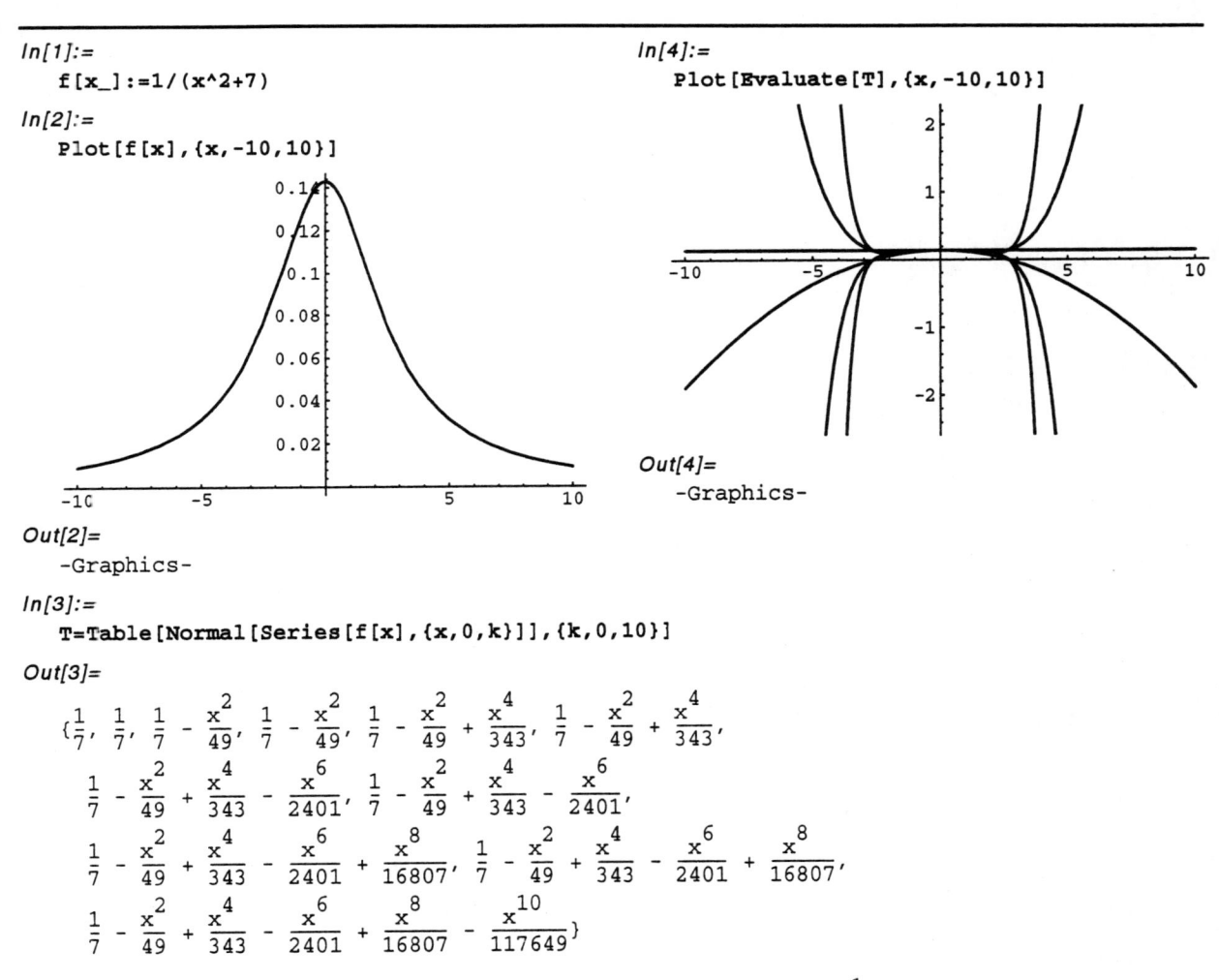

Out[4]=
  -Graphics-

Figure 10.3: **Taylor polynomials of** $\dfrac{1}{x^2+7}$

From the picture, the Taylor polynomials appear to stay close to $\dfrac{1}{x^2+7}$ from about $x = -2.5$ to $x = 2.5$. Outside this interval the separation in the graphs is apparent. Thus, the interval of convergence is approximately $(-2.5, 2.5)$. We emphasize that it is not possible to determine this interval exactly by looking at graphs; we can only estimate it.

Laboratory Exercise 10.3

Interval of Convergence (CCH Text 10.2)

Name _____   Due Date _____

1. Use graphs as we did in Solved Problem 10.2 to estimate the interval of convergence of the Taylor series of $\ln x$ about $x = 2$, $x = 3$, and $x = 10$.

2. Based on your work in Part 1, conjecture the interval of convergence of the Taylor series of $\ln x$ about $x = n$ for $n > 0$.

3. Use graphs as we did in Solved Problem 10.2 to estimate the interval of convergence of the Taylor series of $\dfrac{1}{1+e^x}$ about $x = 0$.

4. Use graphs as we did in Solved Problem 10.2 to estimate the interval of convergence of the Taylor series of the function $f(x) = \begin{cases} e^{-\frac{1}{x^2}} & \text{if } x \neq 0 \\ 0 & \text{if } x = 0 \end{cases}$ about $x = 0$.

(Remark: To enter this function into *MATHEMATICA*, just evaluate f[x_]:= E^-(1/x^2), and *MATHEMATICA* will treat it as if the function value at $x = 0$ is 0. This example requires some careful thought in order to understand what's going on.)

Laboratory Exercise 10.4

Approximating π (CCH Text 10.3)

Name _____   Due Date _____

Example 4 of Section 10.3 of the CCH Text, explains how to use the Taylor series for the arctangent function to obtain an approximation for π. It is noted there that while this method is very elegant, it is not practical for obtaining many digits of π. A better way, presented in Exercise 19, is to use Machin's formula:

$$\frac{\pi}{4} = 4\arctan(\tfrac{1}{5}) - \arctan(\tfrac{1}{239})$$

Thus if $P(x)$ is the Taylor polynomial of degree n for $\arctan x$, then

$$\pi \approx 16 P(\tfrac{1}{5}) - 4 P(\tfrac{1}{239})$$

1. Obtain the Taylor polynomial of degree 29 for $\arctan x$.

2. Use your answer in Part 1 to approximate π.

277

3. How many digits of accuracy did you get from your answer in Part 2? (<u>Hint</u>: Evaluate **N[Pi - %n]**, where n is the line number containing your approximation for π.)

4. Let n be an odd integer. It can be shown that if $P(x)$ is the Taylor polynomial of degree n for arctan x, then the error in approximating π using this method is no more than

$$\frac{4}{n+2}\left(\frac{4}{5^{n+2}} - \frac{1}{239^{n+2}}\right)$$

What degree Taylor polynomial is needed to obtain 100 digits of π using this method?

Solved Problem 10.3: The error in Taylor approximations (CCH Text 10.4)

Use the Taylor polynomial of degree 5 about $x = 0$ to approximate $\tan\frac{1}{2}$, and analyze the error.

Solution: Let **TP= Normal[Series[Tan[x],{x,0,5}]]**. We obtain the result in Out[7] of Figure 10.4. We now evaluate **TP/.x->1/2** and approximate it by evaluating **N[%]** to get 0.545833 in Out[8].

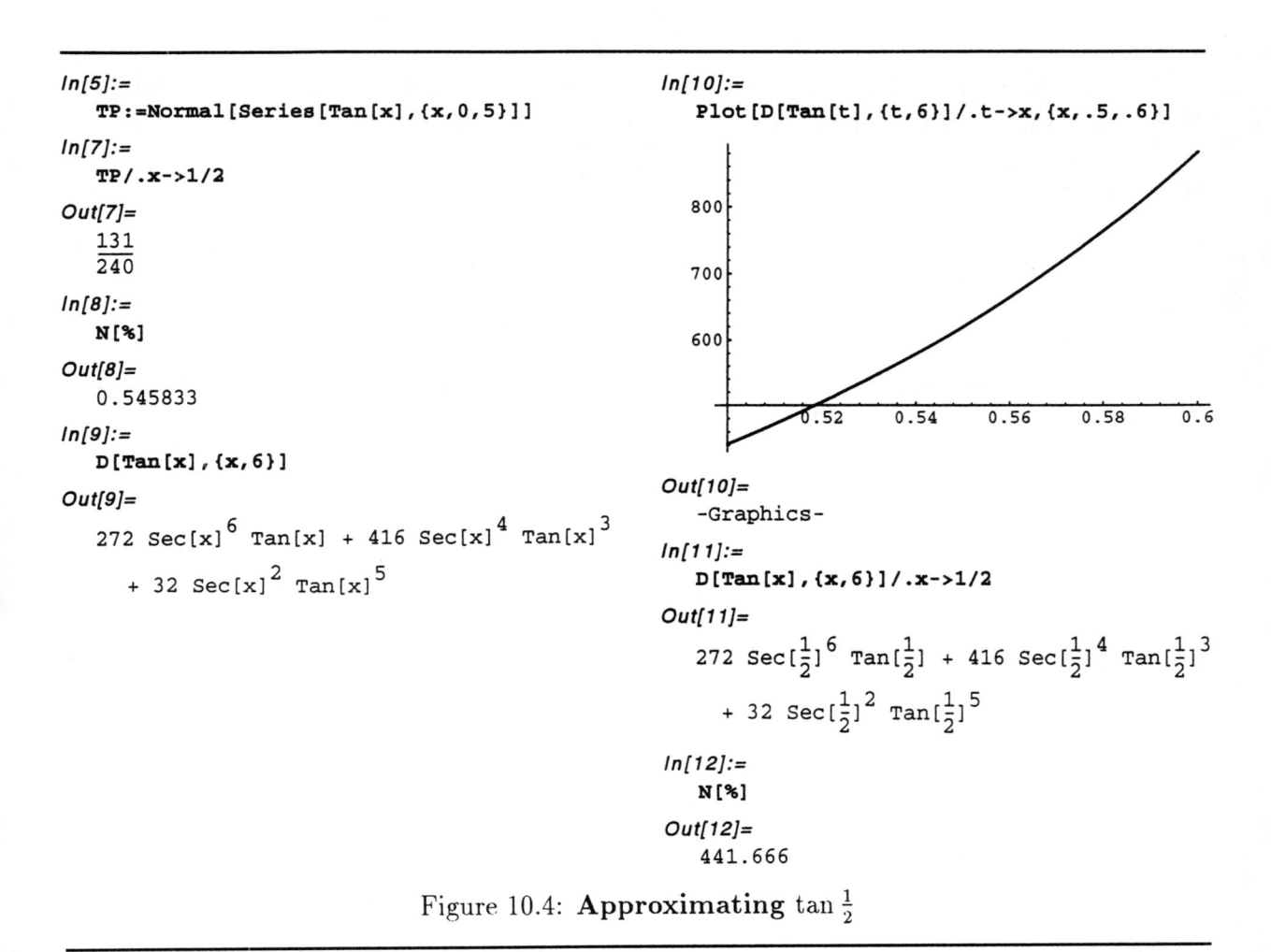

```
In[5]:=
   TP:=Normal[Series[Tan[x],{x,0,5}]]
In[7]:=
   TP/.x->1/2
Out[7]=
   131
   ---
   240
In[8]:=
   N[%]
Out[8]=
   0.545833
In[9]:=
   D[Tan[x],{x,6}]
Out[9]=
   272 Sec[x]^6 Tan[x] + 416 Sec[x]^4 Tan[x]^3
    + 32 Sec[x]^2 Tan[x]^5
```

```
In[10]:=
   Plot[D[Tan[t],{t,6}]/.t->x,{x,.5,.6}]
```

[Graph showing curve rising from about 500 at x=0.52 to above 800 at x=0.6, with gridlines at 600, 700, 800 on y-axis and 0.52, 0.54, 0.56, 0.58, 0.6 on x-axis]

```
Out[10]=
   -Graphics-
In[11]:=
   D[Tan[x],{x,6}]/.x->1/2
Out[11]=
   272 Sec[1/2]^6 Tan[1/2] + 416 Sec[1/2]^4 Tan[1/2]^3
    + 32 Sec[1/2]^2 Tan[1/2]^5
In[12]:=
   N[%]
Out[12]=
   441.666
```

Figure 10.4: **Approximating** $\tan\frac{1}{2}$

The maximum possible error in this approximation is determined by using the error bound formula in Section 10.5 of the CCH Text. In this case the error is no more than $\dfrac{M}{6!}\left(\dfrac{1}{2}\right)^6$, where M is the maximum value of the sixth derivative of the tangent function on the interval

$[0, \frac{1}{2}]$. To find out what M is, let's look at the graph of the sixth derivative of $\tan x$. Evaluate **D[Tan[x],{x,6}]** to see the 6th derivative. Evaluate **Plot[D[Tan[t],{t,6}]/.t->x,{x,.5,.6}]** to see the graph in Out[10] of Figure 10.4. Since this graph is increasing, it attains its maximum value on the interval $[0, \frac{1}{2}]$ at the right-hand endpoint, $\frac{1}{2}$. Evaluate **D[Tan[x],{x,6}]/.x->1/2**, numerically approximate, and from Out[12] we see that we see that $M \approx 441.666$. Therefore, the error is no more than

$$\frac{M}{6!}\left(\frac{1}{2}\right)^6 \approx \frac{441.665}{6!}\left(\frac{1}{2}\right)^6 \approx 0.00958474$$

If we were to evaluate **Tan[1/2] - 0.545833**, we would see that the actual error is 0.00046949. Notice that this is smaller than the predicted error, 0.0095847. This illustrates the fact that $\frac{M}{(n+1)!}x^{n+1}$ gives an upper bound for the error, not the exact value of the error.

Laboratory Exercise 10.5

Approximating $\sec \frac{1}{2}$ (CCH Text 10.5)

Name _____   Due Date _____

1. Use the fifth-order Taylor polynomial for $\sec x$ about $x = 0$ to approximate $\sec \frac{1}{2}$.

2. Find the maximum error in your approximation above. Explain how you made your choice of M in the error formula.

3. Evaluate **Sec[1/2]**. Compare this with the value in Part 1.

4. What is the smallest degree Taylor polynomial about $x = 0$ for $\sec x$ that can be used to approximate $\sec(\frac{1}{2})$ to 4 digits accuracy?

Solved Problem 10.4: Fourier series (CCH Text 10.5)

Find the Fourier approximations of orders 1 through 4 for a periodic function that agrees with x on the interval $[-\pi, \pi]$. Graph $y = x$ together with the Fourier approximation.

Solution: It is easier to type some general code that calculates Fourier series for an arbitrary function than it is to do all the work for a single example; so, that is what we will do. Begin a fresh *MATHEMATICA* notebook and type in each of the following expressions on separate lines exactly as they appear below.

```
Clear[a,b,c,d,s,fs]

s = Sin[k*x]

c = Cos[k*x]

a[y_]:= 1/(2Pi)*Integrate[y,{x,-Pi,Pi}]

b[y_]:= 1/Pi Integrate[y*s,{x,-Pi,Pi}]

d[y_]:= 1/Pi Integrate[y*c,{x,-Pi,Pi}]

fs[y_,n_]:= a[y]+ Sum[s*b[y]+ c*d[y],{k,1,n}]
```

You may save the file for later use by clicking the mouse on the File icon and selecting saveAs. Name the file FOURIER.MA and save it. This will allow you to open the file in the future without having to retype the above code. The function $fs(y, n)$ produces the nth Fourier approximation for the periodic function that agrees with $y(x)$ on the interval $[-\pi, \pi]$.

In our case, $y(x) = x$. Thus we evaluate **Clear[T]**, **T = {x,fs[x,1],fs[x,2],fs[x,3],fs[x,4]}** and **Plot[Evaluate[T],{x,-2,2}]**. The graphs appear in Out[14] of Figure 10.5.

In[1]:=
```
Clear[a,b,c,d,s,fs]
```
In[2]:=
```
s=Sin[k*x]
```
Out[2]=
Sin[k x]

In[3]:=
```
c=Cos[k*x]
```
Out[3]=
Cos[k x]

In[4]:=
```
a[y_]:= 1/(2Pi)Integrate[y,{x,-Pi,Pi}]
```
In[5]:=
```
b[y_]:= 1/(2Pi)Integrate[s*y,{x,-Pi,Pi}]
```
In[6]:=
```
d[y_]:= 1/(2Pi)Integrate[c*y,{x,-Pi,Pi}]
```
In[7]:=
```
fs[y_,n_]:= a[y] + Sum[s*b[y]+c*d[y],{k,1,n}]
```
In[10]:=
```
T={x,fs[x,1],fs[x,2],fs[x,3],fs[x,4]}
```
Out[10]=

$$\{x,\ \text{Sin}[x],\ \text{Sin}[x] - \frac{\text{Sin}[2\ x]}{2},\ \text{Sin}[x] - \frac{\text{Sin}[2\ x]}{2} + \frac{\text{Sin}[3\ x]}{3},$$
$$\text{Sin}[x] - \frac{\text{Sin}[2\ x]}{2} + \frac{\text{Sin}[3\ x]}{3} - \frac{\text{Sin}[4\ x]}{4}\}$$

In[14]:=
```
Plot[Evaluate[T],{x,-7,7},PlotRange->{-2,2}]
```

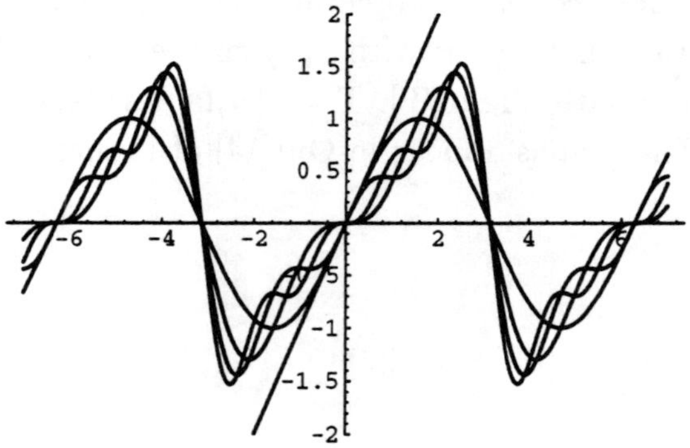

Out[14]=
-Graphics-

Figure 10.5: **Fourier approximations for** $y = x$

Laboratory Exercise 10.6

Fourier Approximations for $|x|$ (CCH Text 10.6)

Name _____   Due Date _____

1. Find the first 5 Fourier approximations to the periodic function that agrees with $|x|$ on the interval $[-\pi, \pi]$.

2. Attach a graph showing $|x|$ and its first 5 Fourier approximations. Explain the behavior of the graphs both on and outside the interval $[-\pi, \pi]$.

Appendix I

MATHEMATICA Version 2.2 Reference and Tutorials

The *MATHEMATICA* program comes with a complete manual, but many of us seem to read software documentation only as a last resort. This appendix is a concise reference to the major features of *MATHEMATICA* that arise often in the exercises in this book. This appendix does not replace *MATHEMATICA: A System for Doing Mathematics by Computer* (Second Edition) for detailed information, and we encourage the interested reader to explore it.

This appendix applies only the current version; namely, version 2.2 for MSDos machines and version 2.2 for Macintosh machines. Older versions may not have all the capabilities mentioned in this appendix.

Let's begin with a brief word about *MATHEMATICA*. There are two main features of the software: the user interface or "front end" and the kernel. Basically, the user interface is how you communicate with *MATHEMATICA*. The kernel is part of *MATHEMATICA* that actually performs the computations. We will not say much about the kernel except that it is basically the same on all machines, from PCs and Macintoshes to workstations and very large computers. The kernel need not even be on the machine you are working on. You will communicate with the kernel via the user interface, which varies from machine to machine. The most common user interface is called the notebook interface. There is a notebook interface for computers running Microsoft Windows ®, version 3.1, and one for the Macintosh line of computers. This tutorial assumes that you have a notebook interface. There are some differences between the Windows notebook and the Macintosh notebook, which we will not elaborate, but you should be able to use this book on either type of computer with no problems.

Starting *MATHEMATICA* varies depending on the setup of your computer. If you have a *MATHEMATICA* folder on Windows, you can open that folder by moving the mouse pointer to the folder icon and double clicking. Then move the pointer to the *MATHEMATICA* icon and double click again. A similar procedure will work on a Macintosh computer. You must determine where the *MATHEMATICA* folder is and open the appropriate folders to get to the *MATHEMATICA* folder. We suggest you open *MATHEMATICA* and follow this appendix by working with *MATHEMATICA* as you read the appendix.

MATHEMATICA can be driven by a menu at the top of the screen. You will see **File Edit Cell Graph Action Style** and other icons, depending on your computer. Select an icon by moving the mouse pointer to the icon and holding down the button. Once you do this, a menu of selection appears. While holding down the button you can move the pointer to the selection you want to activate. Highlighting the selection and then releasing the mouse button will activate the selection. For example, to end your *MATHEMATICA* session, move the mouse to the **File** icon, hold down the button, and drag the mouse pointer to **Quit** at the bottom of the menu of selections. Once **Quit** is highlighted, releasing the button activates the command, which, in this case, closes *MATHEMATICA*. On both a PC and a Macintosh there are keystrokes for many of the commands and we leave it to you to decide which you prefer – using the mouse or typing in commands. Since the commands differ on the two systems, we will restrict ourselves to use of the mouse.

Entering an expression involves customary syntax: addition (+), subtraction (−), division (/), exponents (∧), and multiplication (∗). (However, multiplication of a constant times a nonconstant does *not* require a ∗; 2x is the same as 2∗x. Be warned that xy is considered as one variable. Use x∗y to multiply x times y.)

If you need help during a session, move the mouse to the **Help** icon at the top of the screen, hold down the button, and move the mouse the the selection you want. Releasing the button, activates that selection. You should practice using the mouse if you are not familiar with it. *MATHEMATICA*'s Help option will provide information on all aspects of the program.

Special Functions and Symbols

E	the number e
Pi	the number π
I	the imaginary number i
Infinity	∞
−Infinity	$-\infty$
Sqrt	square root
Abs	absolute value

Section 1. Basics Tutorial

Moving to the mouse to the notebook and clicking once activates the notebook. Type the expression **x^2+3x-4**. To have *MATHEMATICA* "evaluate" this entry, you can either type Shift-Enter (that is, press down the Shift and Enter keys at the same time) or move the mouse to the Action icon at the top of the screen, depress the button, drag down until Evaluate Selection is highlighted, and then release the button. (On a Macintosh, either use Shift-Return or use the Enter key on the far right of the keyboard.) HENCEFORTH, WE WILL JUST SAY "EVALUATE" FOR ANY OF THESE OPTIONS. The first time you evaluate an entry, *MATHEMATICA* loads the kernel, which takes a few seconds. You should see In[1] and Out[1] of Figure 0.1. Once you have evaluated an input, *MATHEMATICA* labels the line you typed as In[n] and its response as Out[n], where n increases as your session progresses. You can refer to these lines later in the session using %n. Furthermore, % by itself refers to the last output.

Thus, evaluating % or %n produces a repetition of the last input line or input n, respectively. See In[2] and Out[2].

You might wonder what happens if you just hit the Enter or Return key. *MATHEMATICA* starts a new line, but it is still part of the SAME entry. Thus, if you are entering a long formula, you can do this on a number of lines. When you evaluate an input that uses more than one line, *MATHEMATICA* ignores the fact you have used more than one line and treats the whole as one input.

MATHEMATICA provides a number of ways to manipulate algebraic expressions. The main ones are Expand, Factor, Simplify, and Together. Expand[*expr*] expands out products and positive integer powers of *expr*. Evaluate **Expand[(x^2-4)^2]** and you should get Out[3] of Figure 0.1. Now evaluate **Factor[%]** to factor a polynomial over the rational numbers. The Simplify function follows rules to "simplify" an expression. The Together function puts an expression over a common denominator.

When using the Expand function, notice the use of the square brackets and that the first letter of the function name is capitalized. This is the standard format in *MATHEMATICA*. Predefined *MATHEMATICA* function names have first letters capitalized and the square brackets are always used. Evaluate **Expand[(x-1)^2]** and you get an error message because the brackets are incorrect. Try it.

In algebraic expressions, use parentheses (and) to group elements. Thus, 4 times 5-3 should be written 4(5-3) (which is 8), whereas 4*5 -3 yields 17. Similarly, $x - y/z$ is the same as $x - (y/z)$ and NOT $(x-y)/z$. You should always be careful that your typed expression is exactly what you want it to be. The correct use of parentheses assures this.

Getting back to our *MATHEMATICA* notebook, evaluate **Factor[%]** and you should see Out[6] of Figure 0.1.

You may name expressions in *MATHEMATICA* using the = sign. Evaluate **f= x^2+4x-6** and then evaluate **f**. See Out[8].

Often you will want to substitute one quantity for another. Suppose we want to find the value of f when $x = 4$. Evaluate **f/.x->4** and you obtain Out[9] of Figure 0.1. The syntax of this is *expr /. a -> b*. This causes *MATHEMATICA* to replace all occurrence of a in *expr* with b. Evaluate **f/.x-> a^3** and see what you get.

When possible, *MATHEMATICA* gives exact answers. For example, evaluate **(34/57)Pi**. To find a decimal approximation evaluate **N[(34/57)Pi]** or **N[%]** as in Out[11]. To get higher degrees of accuracy use N[expr,n], which gives *expr* to n decimal places. Evaluate **N[Pi,20]** and you get π to 20 decimal places.

```
In[1]:=                    In[5]:=                    In[9]:=
    x^2+3x-4                   Expand[(x-1)^2]            f/.x->4
Out[1]=                    Out[5]=                    Out[9]=
    -4 + 3 x + x^2             1 - 2 x + x^2              26
In[2]:=                    In[6]:=                    In[10]:=
    %1                         Factor[%]                  (34/57)Pi
Out[2]=                    Out[6]=                    Out[10]=
    -4 + 3 x + x^2             (-1 + x)^2                 34 Pi
In[3]:=                    In[7]:=                        -----
    Expand[(x^2-4)^2]          f=x^2+4x-6                  57
Out[3]=                    Out[7]=                    In[11]:=
    16 - 8 x^2 + x^4           -6 + 4 x + x^2             N[%]
In[4]:=                    In[8]:=                    Out[11]=
    Factor[%]                  f                          1.87393
Out[4]=                    Out[8]=                    In[12]:=
    (-2 + x)^2 (2 + x)^2       -6 + 4 x + x^2             N[Pi,20]
                                                      Out[12]=
                                                          3.14159265358979323846
```

Figure 0.1: **Algebra tutorial**

Section 2. Syntax

Before getting to more specific *MATHEMATICA* techniques, let's discuss some general features of the program. The point of this section is to alert you about things to watch out for. One of the most difficult points beginners have with *MATHEMATICA* is its syntax. Should you use {, or [or (? When must you use =, ==, or := ? There are some general rules that should help reduce syntactical frustrations. If you are aware of these, you can avoid some of the pitfalls in learning to use *MATHEMATICA*. As a general rule, look at the SYNTAX of all the examples carefully and be sure you type exactly what is shown.

Predefined *MATHEMATICA* functions have the syntax FunctionName[ ]. Note the capital letters and use of the square brackets. In general, *MATHEMATICA* function names have the first letter of words capitalized and the input of the function is enclosed in square brackets. The basic data structure in *MATHEMATICA* is the list. Lists have the form: {entry1,entry2,...}. Notice the use of the set brackets and that entries are separated by commas. *MATHEMATICA* uses the set brackets to collect things together in a structure called a *list*. For a more detailed explanation of lists, we refer you to the *MATHEMATICA* manual, Part 1, 1.8. The main point here is that you should examine how the set brackets are used in the examples that follow in this Appendix. The parentheses (and) are used as they are in mathematics, that is, to block operations in the correct order.

For example, {1,2,{3}} is a list with three elements, 1, 2, and {3}. In *MATHEMATICA* **Sin[x]** is the proper syntax for $\sin x$. You would use **(x-y)/3-x** to mean $\frac{x-y}{3} - x$, whereas **(x-y)/(3-x)** would mean $\frac{x-y}{3-x}$.

MATHEMATICA employs three types of equality signs: =, ==, and := . The sign = means the left-hand side is defined to be the right-hand side. So **g = 3** defines g to be the number 3. Thereafter, g will be the same as 3. The sign == is a logical test. **f[x] == 3** does not change the value of $f[x]$. The double equal sign is used in functions such as Solve, NSolve, and FindRoot, which will be discussed in Sections 7 and 8 below. Finally, the sign := is used in function definitions that you create. You can define functions; for example, **f[x]:= x^2** defines the function $f[x]$ to be x^2. We discuss this more fully in Section 5 below.

Once something is defined in *MATHEMATICA*, the definition stays in force until you quit *MATHEMATICA*. You should get into the habit of making sure no previous definition is in effect when you create a definition. The *MATHEMATICA* function Clear does this for you. Thus **Clear[f,a]** has the effect of removing any definitions attached to f and to a. We will use Clear[ ] often. Note the capital C and the square brackets.

Section 3. Plotting Graphs

To plot a graph, use *MATHEMATICA*'s Plot function. Its basic syntax is Plot[*expr*,{x,a,b}] which causes *MATHEMATICA* to plot *expr* (which should be a function of x) from $x = a$ to $x = b$. Evaluate **Plot[x^3-2x^2 +3,{x,-3,3}]**. You should get the graph above Out[1] of Figure 0.2.

Note that *MATHEMATICA* automatically determined the range and the location of the axis. You can control the range plotted by adding PlotRange->{c,d}. Evaluating **Plot[x^3-2x^2+3,{x,-3,3},PlotRange- >{2,4}]** yields the graph above Out[2]. You can control where the axes cross with AxesOrigin->{s,t}. We evaluated **Plot[x^3-2x^2+3,{x,-3,3},AxesOrigin->{1,1}]** in Out[3].

Given a plot, you can change its size by moving the mouse pointer into the plot and clicking the mouse button. *MATHEMATICA* puts a border around the plot. Move the mouse pointer to one of the rectangles on the border. When the pointer becomes <->, you can enlarge or shrink the graphics window by holding down the button, dragging the mouse, and then releasing the button. Try it on a graphics window you have created. Another thing you can do is find the coordinates of points in the graphics window. Move the mouse pointer to a position in the graphics window and hold down the Ctrl key on a PC or the Command and Option keys on

In[1]:=
    Plot[x^3-2x^2+3,{x,-3,3}]

Out[1]=
    -Graphics-

In[2]:=
    Plot[x^3-2x^2+3,{x,-3,3},PlotRange->{2,4}]

Out[2]=
    -Graphics-

In[3]:=
    Plot[x^3-2x^2+3,{x,-3,3},AxesOrigin->{1,1}]

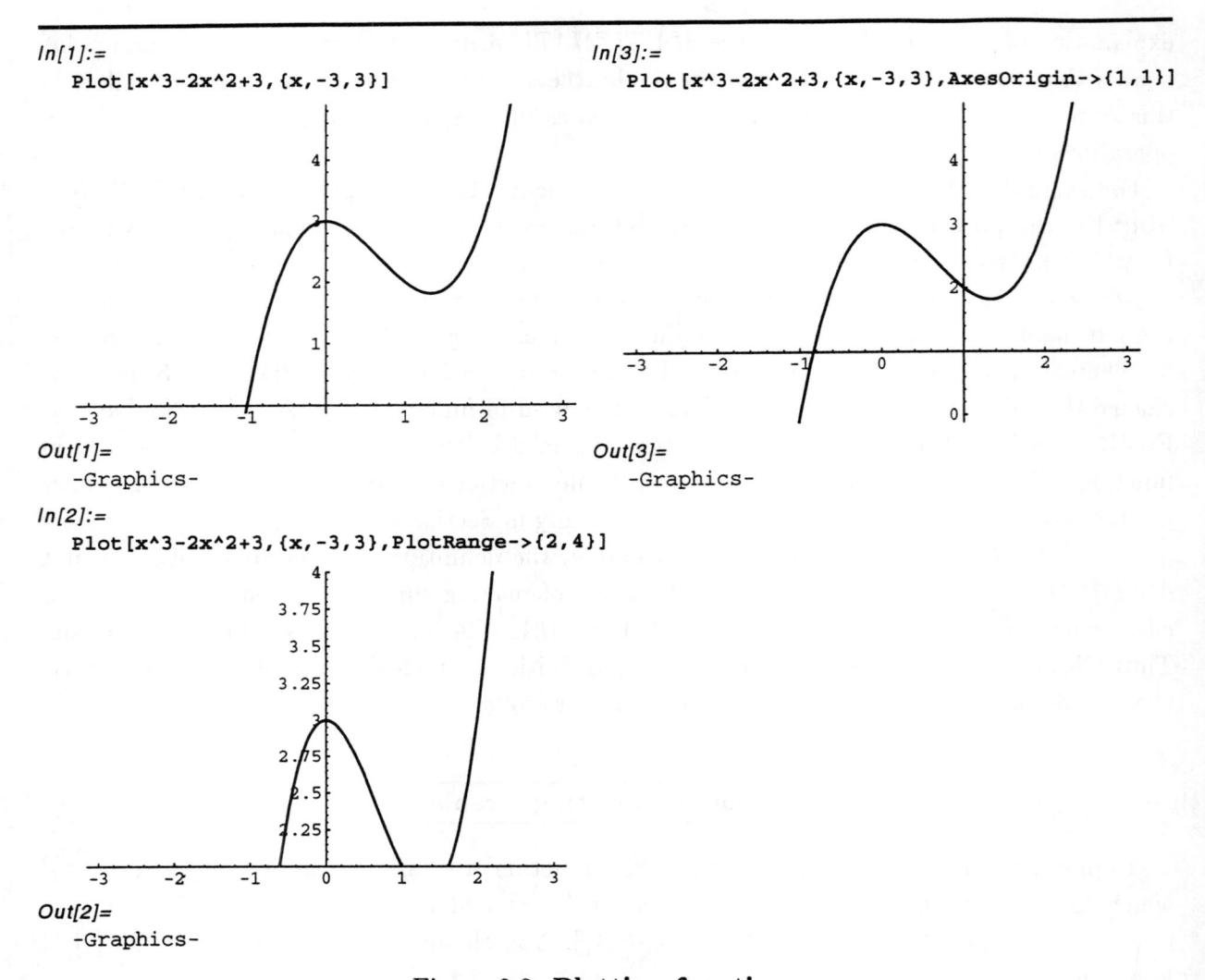

Out[3]=
    -Graphics-

Figure 0.2: **Plotting functions**

a Macintosh. In the left-hand side of the bottom margin, the coordinates of the point are displayed. You should try this and experiment. Hold down the keys as you move the mouse.

> **Summary of Plot Features**
>
> - Plot[fn,{x,a,b}] plots a function fn from $x = a$ to $x = b$.
>
> - Adding PlotRange->{c,d} restricts the range from $y = c$ to $y = d$.
>
> - Adding AxesOrigin->{s,t} causes the axes to cross at (s, t).
>
> - To change the size of a graphics window, click the mouse while the pointer is in a graphics window. Move the pointer to one of the rectangles in the border of the window until the pointer changes to <->. Holding the mouse button, resize the window by moving the mouse and then releasing the button.
>
> - After clicking in a graphics window, holding down the Ctrl key on a PC or the Command-Option keys on a Macintosh permits one to see the coordinates of the cursor in the left-hand corner of the bottom border of the *MATHEMATICA* window.

Plotting several graphs at once: If you want to plot two or more graphs, you can put them in a list separated by commas and enclosed in set brackets. For example, let's plot $\sin x$, $\sin(2x)$, and $\sin(3x)$ from $x = -2\pi$ to $x = 2\pi$. Evaluate **Plot[{Sin[x],Sin[2x],Sin[3x]}, {x,-2Pi,2Pi}]**. (See Out[2] of Figure 0.3.) It is hard to distinguish one graph from another. To help with this, you can change the thickness of the graphs. The syntax of the command you add is PlotStyle->{Thickness[n1],Thickness[n2],Thickness[n3]}. Evaluate **Plot[{ Sin[x], Sin[2x], Sin[3x] },{ x, -2Pi, 2Pi }], PlotStyle -> { Thickness[0.001], Thickness[0.006],Thickness[0.009]}]**. See In[3] and Out[3] of Figure 0.3. Note that we used three lines for the input. Remember, you can enter this command on more than one line if you wish. You should experiment with different values for the thickness to get output you like. If you want just two graphs, for example, $\sin x$ and $\sin(2x)$, you evaluate **Plot[{Sin[x],Sin[2x],{x,-2Pi,2Pi},PlotStyle->{ Thickness[0.001],Thickness[0.007]}]**. Try it.

Color monitors: If you have a color monitor, you can use different colors instead of different thicknesses. The syntax for this is similar:

$$\text{PlotStyle->}\{\text{RGBColor[n1,n2,n2], RGBColor[m1,m2,m3],...}\}$$

Here $n1$, $n2$ and $n3$ are numbers between 0 and 1. The first gives the red value, the second the green value, and the third the blue value. Thus to get an all green plot, use

In[2]:=
Plot[{Sin[x],Sin[2x],Sin[3x]},{x,-2Pi,2Pi}]

Out[2]=
-Graphics-

In[3]:=
**Plot[{Sin[x],Sin[2x],Sin[3x]},{x,-2Pi,2Pi},
  PlotStyle->{Thickness[0.001],
  Thickness[0.006], Thickness[0.009]}]**

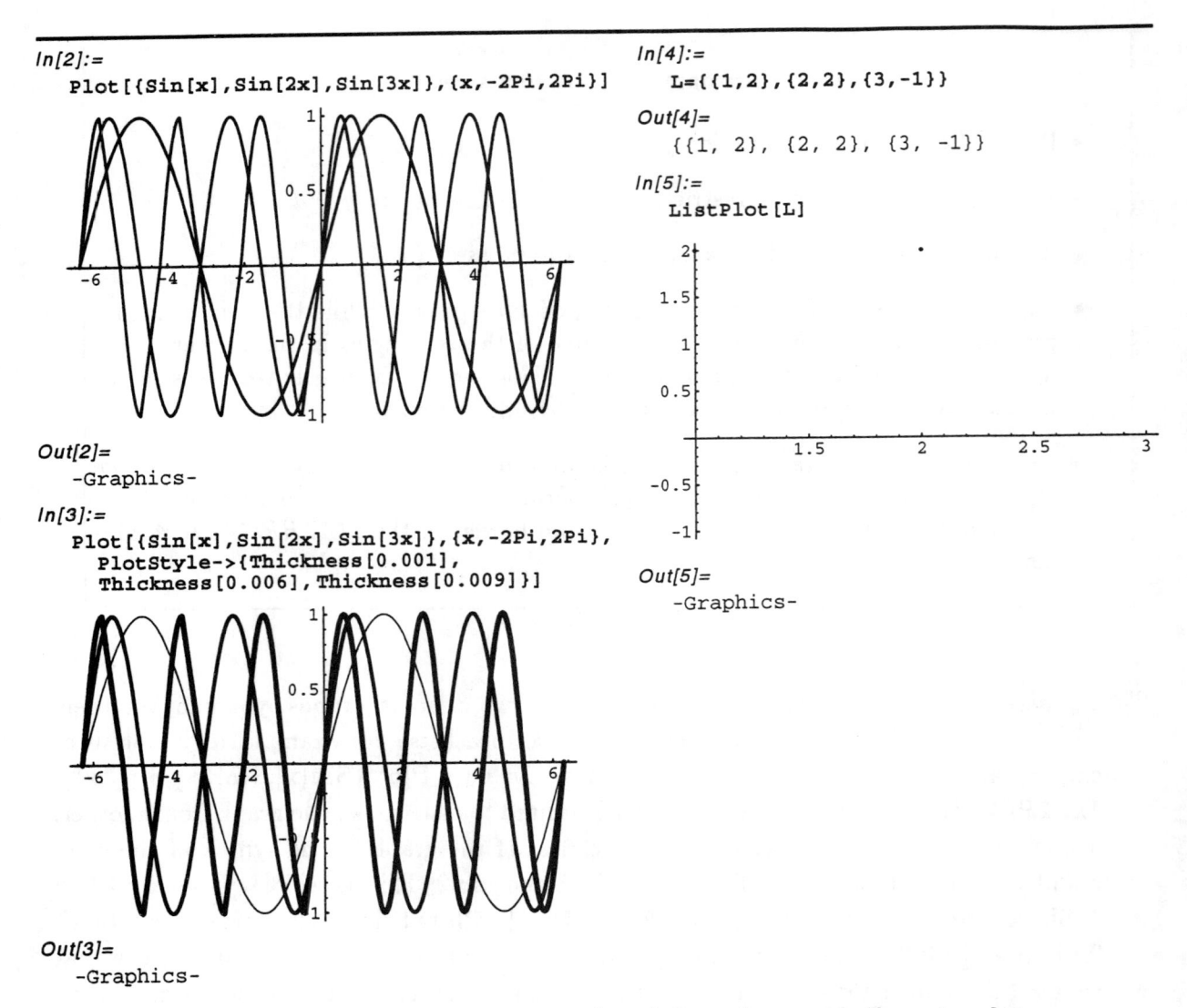

Out[3]=
-Graphics-

In[4]:=
L={{1,2},{2,2},{3,-1}}

Out[4]=
{{1, 2}, {2, 2}, {3, -1}}

In[5]:=
ListPlot[L]

Out[5]=
-Graphics-

Figure 0.3: **Multiple graphs, restricted domains, and discrete plots**

RGBColor[0,1,0]. Evaluate **Plot[{ Sin[x], Sin[2x] }, { x, -2Pi, 2Pi }, PlotStyle -> {RGBColor[1,0,0], RGBColor[0,1,0]}]**. We will not assume that you have a color monitor but if you do, you should experiment with the RGBColor option.

Plotting individual points: Suppose you want to plot several points on the screen, for example, $(1, 2)$, $(2, 2)$, and $(3, -1)$. Evaluate **L = { {1,2},{2,2},{3,-1}}]**. Notice the use of set brackets and commas. This creates a list of points which appears in Out[4] of Figure 0.3. To plot points we use *MATHEMATICA's* ListPlot command whose syntax is ListPlot[*list of points*]. Evaluate **ListPlot[L]** to get the graphics window after In[5].

294

```
In[1]:=
  <<Graphics`Graphics`
In[3]:=
  PolarPlot[Sin[3t],{t,-Pi,Pi}]
```

```
In[5]:=
  ParametricPlot[{Sin[t],Sin[2t]},{t,-Pi,Pi}]
```

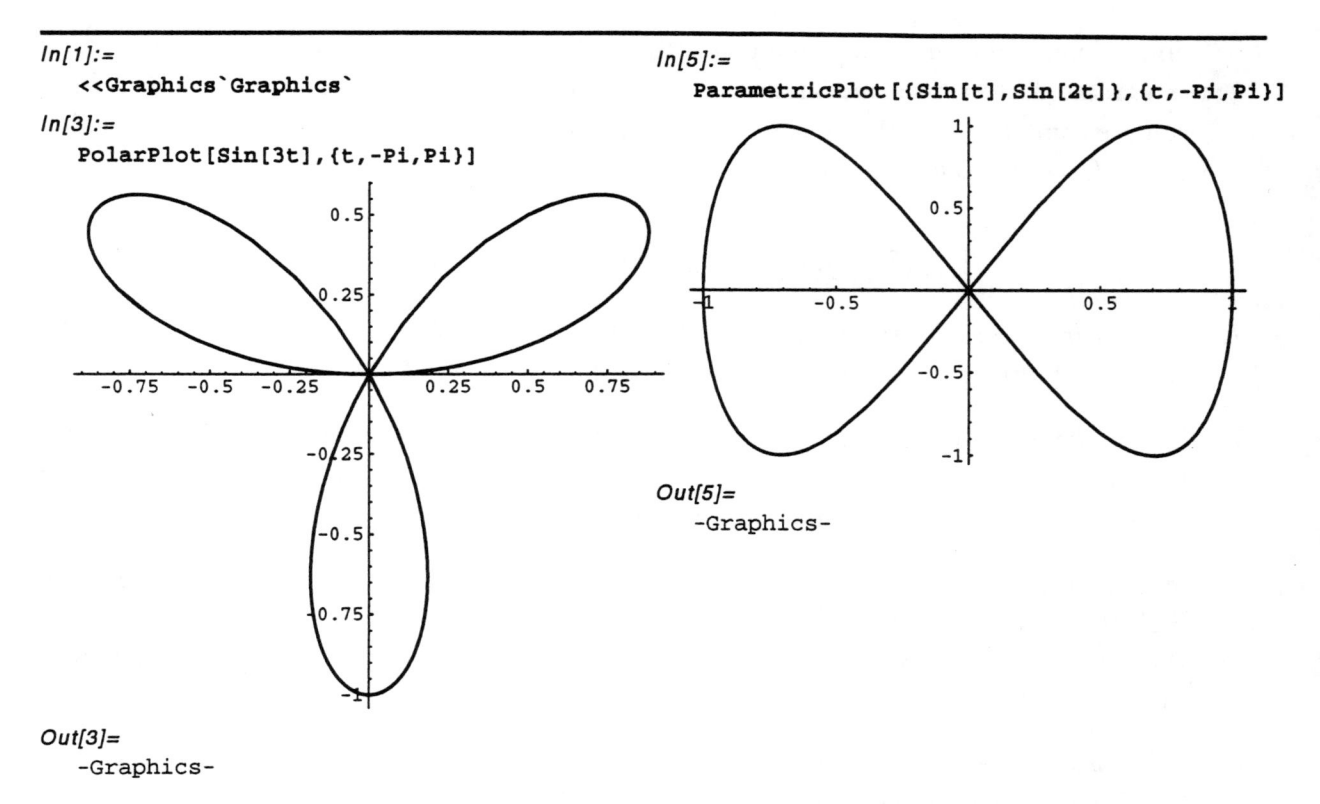

Out[5]=
  -Graphics-

Out[3]=
  -Graphics-

Figure 0.4: **Polar, parametric, and piecewise plots**

Plotting in polar coordinates: Let's plot the polar equation $r = \sin(3\theta)$. We must load a *MATHEMATICA* package for this. Evaluate `<< Graphics`Graphics`". This loads the graphics package that contains the function PolarPlot. The syntax of PolarPlot is PolarPlot[fn,{t,a,b}] where fn is a function of the form $r = f(t)$ with t in radians, and $r = f(t)$ is plotted from $t = a$ to $t = b$. After you have loaded the graphics package, evaluate **PolarPlot[Sin[3t],{t,-Pi,Pi}]** and you get the "three-leafed rose" in the graphics window after In[3] of Figure 0.4.

Plotting parametric curves: Let's plot the graph of the parametric curve, $x = \sin t$, $y = \sin(2t)$ on the interval $[-\pi, \pi]$. We use *MATHEMATICA*'s ParametricPlot function which has syntax ParametricPlot[{fx,fy},{t,a,b}], where fx is the x-coordinate function in variable t and fy is the y-coordinate function in variable t. The command plots the functions as t varies from a to b. Evaluate **ParametricPlot[{Sin[t],Sin[2t]},{t,-Pi,Pi}]**. The graph is displayed after In[5] of Figure 0.4.

Other features: One of *MATHEMATICA*'s most powerful features is its graphics capability. We have just touched the surface of what *MATHEMATICA* can do. There are three dimensional plots and many commands to yield graphics output in a form you want. You can create dashed lines and shaded areas. You can name the axes and much more. Once you are comfortable with the basic *MATHEMATICA* commands, you should read Section 1.9 on Graphics and Sound in the manual.

Four Common Questions About Plotting

The axes of the graph do not cross at $(0,0)$. What's wrong? You can force this to occur by using the plotting option AxesOrigin->{0,0}. See the discussion above.

The values displayed on the axes overlap each other and cannot be read. What can I do? Enlarge the size of the graphics window using the command described above. When you do this, axes will be larger and the values that *MATHEMATICA* enters will not overlap.

How do I plot a vertical line? The easiest way to do this is to load the Graphics package by evaluating << Graphics'Graphics'. Then, to plot the line $x = 3$, for example, evaluate **ParametricPlot[{3,y},{y,-2,2}]**. Another way is described in this manual in Solved Problem on Rational Functions.

I keep getting error messages when I try to plot. Why? You probably have the syntax wrong. Check your brackets and equality signs!

Section 4. Plotting Tutorial

We will use some basic graphing skills to estimate the solutions of $x^5 + 5x^2 - 1 = 0$. Evaluate **Clear[f]** and **f= x^5+5x^2-1** as in inputs 1 and 2 of Figure 0.5. Now we evaluate **Plot[f,{x,-5,5}]** You see the graph displayed after In[4] in Figure 0.5.

The solutions of the equation occur where the graph of $x^5 + 5x^2 - 1$ crosses the x axis, and from Figure 0.5 it is clear that there are 3 solutions occurring somewhere in the interval $[-2, .7]$ Replot the graph using that as the domain.

Let's go after the positive solution first. Place the mouse pointer inside the graphics window, click once, move to one of the boxes on the border of the window, and hold down the mouse button to enlarge the window. Now move the mouse to the leftmost crossing and click once, and hold down either the Ctrl key or the Command-Option combination. Graphically we see that the x value is approximately -1.667. The other two crossings are approximately -0.447 and 0.426. (By replotting with the smaller domains containing a zero, we could get more accurate estimates.) In Section 8 below, you will find a method to obtain even better approximations of these roots.

In[1]:=
   Clear[f]

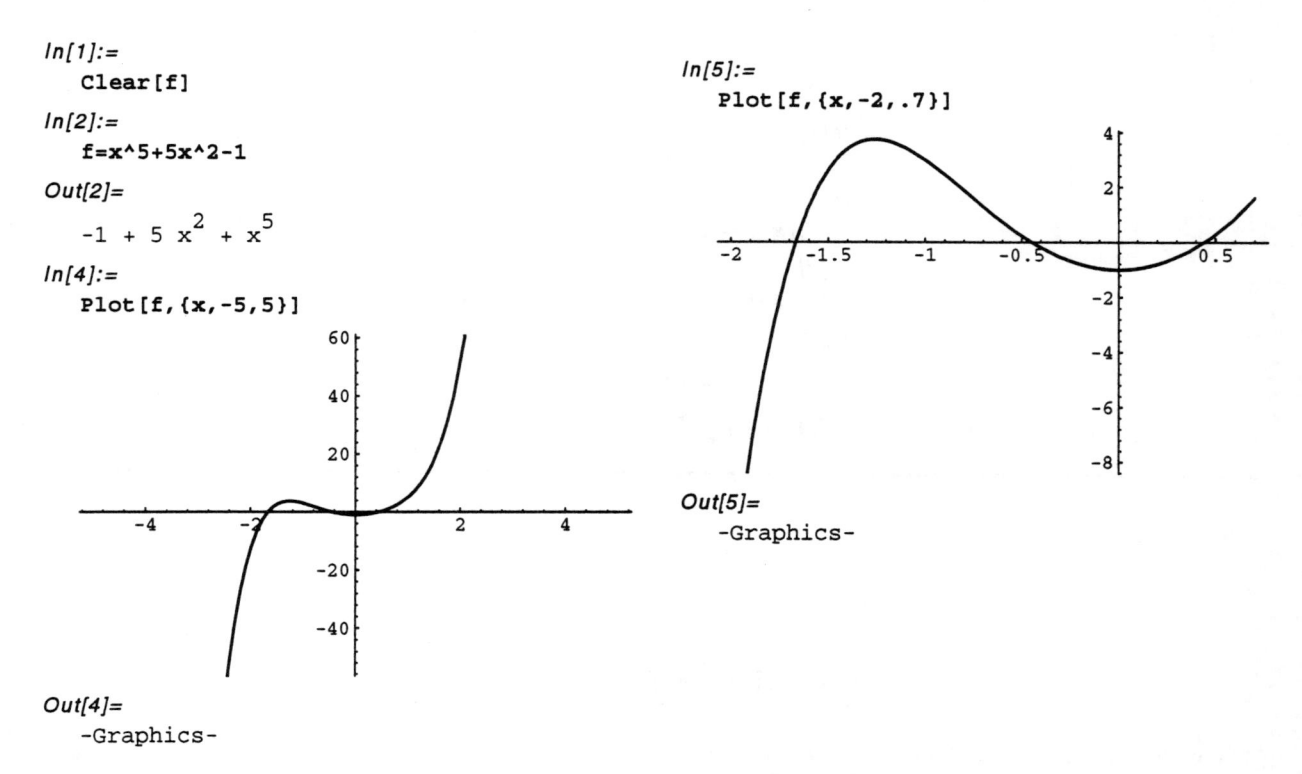

In[2]:=
   f=x^5+5x^2-1

Out[2]=
   $-1 + 5 x^2 + x^5$

In[4]:=
   Plot[f,{x,-5,5}]

In[5]:=
   Plot[f,{x,-2,.7}]

Out[5]=
   -Graphics-

Out[4]=
   -Graphics-

Figure 0.5: **The graph of** $x^5 + 5x^2 - 1$

Section 5. Defining Functions and Constants

To define a function, for example, $f(x) = 1 + x^2$, evaluate **Clear[f]** (to remove any previous definitions) and **f[x_]:= 1+x^2**, as seen in In[2] of Figure 0.6. (Be sure to put the colon before the equal sign and to use square brackets and the underline _ after the variable on the left-hand side. This tells *MATHEMATICA* that f is a function and x is a variable.) Until you quit *MATHEMATICA* or clear f, *MATHEMATICA* will contain the definition $f(x) = 1 + x^2$.

If you want to evaluate the function when $x = 3 + t$, evaluate **f[3+t]**. The result appears in Out[3].

If you want to redefine f or use f in some other way, remember that you must evaluate **Clear[f]** to remove the definition before using f some other way.

If you want to assign the value 2 to the letter a, evaluate **Clear[a]** and **a=2** as seen in inputs 4 and 5 of Figure 0.6. When you evaluate **a^2**, *MATHEMATICA* reports the answer 4. If you want to redefine a, remember to clear the value of a first.

```
In[1]:=                In[4]:=
    Clear[f]               a=2
In[2]:=                Out[4]=
    f[x_]:=1+x^2           2
In[3]:=                In[5]:=
    f[3+t]                 a^2
Out[3]=                Out[5]=
               2           4
    1 + (3 + t)
```

Figure 0.6: **Defining functions and constants**

Section 6. Moving Around and Editing

You can move up and down to see earlier work by placing the mouse pointer on one of the arrows in the right border of the notebook and clicking the mouse button. You can also use the PageUp and PageDown keys.

To use previously defined input, find its number, say n. Then typing %n is the same as retyping that input. You can create new inputs anywhere, and *MATHEMATICA* continues its numbering scheme. You can also move back to old inputs; click the mouse on input line. It is then ready to be edited. When you reevaluate the edited input, *MATHEMATICA* treats it in the same way as if you typed in the whole line (or lines).

In editing, you can use arrow keys to move the cursor, and the delete key or backspace to erase characters. Your Insert key toggles between typeover and insert modes.

Section 7. Solving Equations Exactly: Solve

We begin with *MATHEMATICA*'s Solve command, which will exactly solve some polynomial equations. The syntax is Solve[lhs == rhs,x] where *lhs* and *rhs* are formulas, and x is the variable that is solved for.

Evaluate the equation 2x^4 + 9x^3 - 27x^2 = 22x - 48. The solutions appear in Out[1] of Figure 0.7.

If you ask *MATHEMATICA* to solve an expression that is not an equation, inequality, or other relation, it returns an error message. One of the most common errors is to use = instead of ==, or to omit the equal signs altogether.

Try to find the roots of $x^2 + 4$. Notice in Out[4] that *MATHEMATICA* works with complex numbers.

In[1]:=
    Solve[2x^4+9x^3-27x^2==22x-48,x]
Out[1]=
    {{x -> -(3/2)}, {x -> 2}, {x -> (-5 - Sqrt[57])/2},
    {x -> (-5 + Sqrt[57])/2}}
In[2]:=
    Solve[{x^2-3x+2==0, x^2==4},x]
Out[2]=
    {{x -> 2}}
In[3]:=
    Solve[x+1==x-1,x]
Out[3]=
    {}

In[4]:=
    Solve[x^2+4==0,x]
Out[4]=
    {{x -> -2 I}, {x -> 2 I}}
In[6]:=
    Solve[x^5+5x^2-1==0,x]
Out[6]=
    {ToRules[Roots[5 x^2 + x^5 == 1, x]]}

Figure 0.7: **Solving equations exactly**

When a solution cannot be found exactly, *MATHEMATICA* simply returns the equation (often with some comments). This occurs with $x^5 + 5x^2 - 1$ in Out[6]. In this case it will be necessary to approximate the solutions. See Section 8 below.

MATHEMATICA's Solve can work on a number of equations and find common solutions. Evaluate **Solve[{x^2-3x+2==0,x^2==4},x]**; you should get Out[2] of Figure 0.7.

If *MATHEMATICA* is able to determine that there are no solutions, it will return { }. In In[3] of Figure 0.7 we have attempted to solve $x + 1 = x - 1$.

Section 8. Approximations and Precision

If you have a value, for example, sin 3, and you want to know that value to 8 decimal places, evaluate **N[Sin[3],8]**. The syntax for N is N[numbr,n], where n is optional and gives the number of decimal place accuracy.

MATHEMATICA provides two methods for approximating solutions, NSolve and FindRoot. NSolve is similar to Solve except that it numerically approximates solutions to polynomial equations. Its syntax is NSolve[lhs == rhs,x,n], where n is optional and gives the number of decimal place accuracy. Evaluate **NSolve[x^2==2, x,10]** and you should get Out[1] of Figure 0.8. Also evaluate **NSolve[x^3 -x+4 == 0, x]** to obtain Out[2]. As with Solve, NSolve can also be used with a system of equations.

MATHEMATICA's FindRoot function uses a version of Newton's method, which is discussed in CCH Chapter 5, Section 5.7. The syntax for FindRoot is FindRoot[lhs==rhs,{x ,a}], where x is the variable to be solved for, and a is a value to start Newton's method. The value a should be close to the solution you want so that Newton's method converges on the solution. If the method fails, you will get an error message informing you that it failed to converge in 15 steps.

Let's look at $x^5 + 5x^2 = 1$. We found in Section 7 that *MATHEMATICA* could not solve this equation exactly, but we did estimate the three solutions graphically in Section 7. We will examine a way to get answers correct to several decimal places.

It will be convenient to think of the equation in the form $x^5 + 5x^2 - 1 = 0$. Thus, evaluate **Clear[f]** and **f = x^5 + 5x^2 - 1**. We now evaluate **FindRoot[f==0,{x,-1.6}]** to get -1.66798. Evaluating **FindRoot[f==0,{x,-0.4}]** we get -0.451384. Evaluation of **FindRoot[f,{x,4.2}]** yields 0.443366 as in outputs 6-9 of Figure 0.8. NSolve will work on $x^5 + 5x^2 - 1$, but we used FindRoot to demonstrate its use.

In summary, for most functions other than polynomials of degree no more than four, we usually need to use FindRoot to calculate approximate solutions. Since FindRoot uses Newton's method, we must give FindRoot a starting value close to the solution. To do this, it is best to plot the graph of the function(s) in question and use this to find the starting value(s).

This technique leads to the following question. If you ask a computer to plot a graph, how can you be sure that the screen displays all the points where the graph crosses the x axis? In general, there is no simple answer to this question, but in the case of polynomials the following theorem can help.

The Confinement Theorem. Let $P(x) = a_n x^n + a_{n-1} x^{n-1} + \cdots + a_1 x + a_0$ be a polynomial with $a_n \neq 0$. Let M denote the largest of $|a_0|, |a_1|, ..., |a_n|$, and let $K = \dfrac{nM}{|a_n|}$. Then $P(x)$ has no zeros outside the interval $[-K, K]$.

Proof: Let $|x| > K$. Then for each $j = 0, 1, \ldots, n-1$ we have $|x| > \dfrac{n|a_j|}{|a_n|}$.

```
In[1]:=
   NSolve[x^2==2,x]
Out[1]=
   {{x -> -1.41421}, {x -> 1.41421}}
In[2]:=
   NSolve[x^3-x+4==0,x]
Out[2]=
   {{x -> -1.79632}, {x -> 0.898161 - 1.19167 I},
    {x -> 0.898161 + 1.19167 I}}
In[3]:=
   Clear[f]
In[6]:=
   f=x^5+5x^2-1
Out[6]=
   -1 + 5 x^2 + x^5
```

```
In[7]:=
   FindRoot[f==0,{x,-1.6}]
Out[7]=
   {x -> -1.66798}
In[8]:=
   FindRoot[f==0,{x,-0.4}]
Out[8]=
   {x -> -0.451384}
In[9]:=
   FindRoot[f==0,{x,4.2}]
Out[9]=
   {x -> 0.443366}
```

Figure 0.8: **NSolve** and approximate zeros of $x^5 + 5x^2 - 1$

Since $K > 1$, $|x|^{n-j} > \dfrac{n|a_j|}{|a_n|}$ also. Multiplying both sides of this last inequality by $\dfrac{|a_n x^j|}{n}$ gives

$$\frac{|a_n x^n|}{n} > |a_j x^j| \text{ for each } j = 0, 1, \ldots, n-1$$

By the triangle inequality, $|\sum_{j=0}^{n} a_j x^j| \geq |a_n x^n| - \sum_{j=0}^{n-1} |a_j x^j|$.

Combining the last two inequalities gives

$$|\sum_{j=0}^{n} a_j x^j| \geq |a_n x^n| - \sum_{j=0}^{n-1} |a_j x^j| > |a_n x^n| - \sum_{j=0}^{n-1} \frac{|a_n x^n|}{n} = 0$$

This completes the proof.

If we apply this theorem to the previous problem, where $P(x) = x^5 + 5x - 1$, we see that $M = 5$ and $K = 25$. Therefore the graph cannot cross the x axis outside the interval $[-25, 25]$. You may plot this expression once more to verify that we have found all the zeros. Evaluate **Plot[x^5,+5x-1,{x,-25,25}]**. Notice that the graph does not help in getting approximate values but, by the theorem, we know all the zeros are shown in the graph. We can narrow the domain so that the graph will yield approximate values. Now we can use FindRoot.

> ## A Strategy for Solving Equations
>
> **Step 1**: Use *MATHEMATICA's* Solve to find possible exact solutions. If all the solutions are found, you do not need to proceed further.
>
> **Step 2**: If exact solutions are not found, evaluate **Clear[f]** and **f[x_] := fn** (or **f = fn**) where $fn = 0$ is what we are trying to solve. Then evaluate **Plot[f[x],{x,a,b}]** and use the mouse to find rough approximations of the zeros. You may have to change the values of a and b a few times to get all the zeros.
>
> **Step 3**: For each rough approximate zero, z, evaluate **FindRoot[f==0,{x,z}]**.
>
> **Step 4**: Be alert to the possibility that zeros may be hiding outside the range of the plot. In the case of a polynomial, you may use the confinement theorem. More generally, you should plot the function using large domains.

Section 9. Calculus

MATHEMATICA contains a number of calculus operations. They include D (for differentiation), Integrate, NIntegrate, Limit, Sum, Product, Series (for Taylor polynomials), and DSolve (for solving differential equations).

Example: Integrate $x^2 + 3x - 4$. Evaluate **Integrate[x^2+3x-4,x]**. See Out[1] in Figure 0.9. To get the definite integral from $x = 3$ to $x = 5$, evaluate **Integrate[x^2+3x-4,{x,3,5}]** to get Out[2]. The syntax for the indefinite integral is Integrate[fn,x] to get $\int fn\ dx$. The syntax for the definite integral is Integrate[fn,{x,a,b}] to get $\int_a^b fn\ dx$.

Example: Integrate $e^{\sin^2 x}$ from 0 to π. This integral can only be approximated. Try to use the Integrate command. On the other hand, evaluate **NIntegrate[E^(Sin[x]^2),{x,0,Pi}]** and we get 5.50843 as in Out[3].

Example: Find the second derivative of $x^3 + 3x - 4$. Evaluate **D[x^3 + 3x -4,{x,2}]** to get the second derivate (with respect to x) as in Out[6] of Figure 0.9. Most calculus books use $\frac{d^2}{dx^2}(x^3 + 3x - 4)$. If we wanted the first derivative, we would evaluate **D[x^3 + 3x -4,x]**. To find the second derivative at $x = 4$, evaluate **D[x^3 + 3x -4,{x,2}]/.x->4** to obtain Out[7]. The syntax is D[fn,{x,n}] for the nth derivative of fn with respect to x.

Example: Find the sum of the series $\sum_{n=1}^{\infty} 3^{-n}$.

```
In[1]:=
    Integrate[x^2+3x-4,x]
Out[1]=
    -4 x + (3 x^2)/2 + x^3/3
In[2]:=
    Integrate[x^2+3x-4,{x,3,5}]
Out[2]=
    146/3
In[3]:=
    NIntegrate[E^(Sin[x]^2),{x,0,Pi}]
Out[3]=
    5.50843
In[6]:=
    D[x^3+3x-4,{x,2}]
Out[6]=
    6 x
In[7]:=
    D[x^3+3x-4,{x,2}]/.x->4
Out[7]=
    24

In[10]:=
    Sum[3^(-n),{n,1,Infinity}]
Out[10]=
    Sum[3^-n, {n, 1, Infinity}]
In[11]:=
    Sum[3^(-n),{n,1,30}]
Out[11]=
    1029455660473 24 / 205891132094649
In[12]:=
    NSum[3^(-n),{n,1,Infinity}]
Out[12]=
    0.5
In[13]:=
    Sum[3^i,{i,1,2,0.25}]
Out[13]=
    27.9829
In[14]:=
    Limit[(1+1/x)^x,x->Infinity]
Out[14]=
    E
In[15]:=
    Normal[Series[Sin[x^2],{x,0,6}]]
Out[15]=
    x^2 - x^6/6
```

Figure 0.9: **Calculus operations**

Evaluate **Sum[3^(-n),{n,1,Infinity}]** and we see that *MATHEMATICA* cannot evaluate this exactly. We can try evaluating **Sum[3^(-n),{n,1,b}]** for larger and larger values of b. See Out[11]. We can use the numerical approximation sum, NSum. Evaluate **NSum[3^(-n),{n,1,Infinity}]** and we see that the answer is 0.5 as shown in Out[12] of Figure 0.9. The syntax of Sum and NSum are the same; namely, Sum[fn,{i,a,b}], where fn is a function with index i which is evaluated from $i = a$ to b. If you want to have a step different than 1, the syntax is Sum[fn,{a,b,s}], where s is the step. For example, evaluating **Sum[3^i,{i,1,2,0.25}]** yields Out[13], which is $3^1 + 3^{0.25} + 3^{0.5} + 3^{0.75} + 3^2$.

<u>Example:</u> Find the limit as x goes to infinity of $\left(1 + \frac{1}{x}\right)^x$. Evaluate **Limit[(1 + 1/x)^x,x->Infinity]** and get e, as in Out[14] of Figure 0.9. The syntax for Limit is Limit[fn,x->a], which yields $\lim_{x \to a} fn$.

303

Example: Find the Taylor series of degree 6 for $\sin x^2$ at $x = 0$. Evaluate **Normal[Series[Sin[x^2],{x,0,6}]]** to get Out[15]. (Leaving out the Normal gives the Taylor series with an approximate remainder term.) The syntax for the Series command is Series[fn,{x,a,n}], which yields the Taylor series of degree n at $x = a$ for the function fn.

Section 10. Substituting into an Expression

Often we want to plug a value of x into a function or substitute one expression for another. To do this we evaluate **expr/.x-> expr2**. This has the effect of replacing every occurrence of x in $expr$ with $expr2$. For examples we will use the expressions $(x + 1)^2 + \sin(x + 1)$ and x^3, so evaluate **(x+1)^2 + Sin[x+1]/.x->(x^3)** and you get Out[11] in Figure 0.10. Evaluate **Expand[%]** to get Out[12].

```
In[11]:=
   (x+1)^2 + Sin[x+1]/. x->(x^3)
Out[11]=
   (1 + x^3)^2 + Sin[1 + x^3]
In[12]:=
   Expand[%]
Out[12]=
   1 + 2 x^3 + x^6 + Sin[1 + x^3]
In[13]:=
   %12/.x->3
Out[13]=
   784 + Sin[28]
```

```
In[14]:=
   N[%]
Out[14]=
   784.271
In[15]:=
   %11/.(1+x^3)->t
Out[15]=
   t^2 + Sin[t]
```

Figure 0.10: **Substituting**

Example: Substitute the number 3 into Out[12] for x.

1. Evaluate **%12/.x->3** to get Out[13].
2. To find a numerical approximation, evaluate **N[%]** to get Out[14].

Example: Substitute t into Out[11] for $x^3 + 1$ *everywhere it occurs* in the formula. Evaluate **%11/.(1+x^3)->t** to get Out[15] in Figure 0.10

Section 11. Capital Letters and Multiletter Names

You may want to designate a variable with capital letters or with a name having more than one letter. Since *MATHEMATICA* uses capital letters for its function names, we suggest that

304

you use small letters. Remember that you can name variables and functions with multiple letters; but if you leave a space, for example, $x\ y$, *MATHEMATICA* interprets this as $x*y$. Similarly, if you want to multiply x times y, you should enter $x*y$.

Section 12. Trigonometry

<u>Degrees Versus Radians</u>: The syntax for the cosine of 30 degrees is `Cos[30 Degree]`. If we just entered `Cos[30]`, *MATHEMATICA* would interpret it as cosine of 30 *radians*. If you evaluate **N[30 Degree]** *MATHEMATICA* gives the (approximate) answer in radians, namely 0.523599.

MATHEMATICA can do some trigonometric manipulations. You can use operations like Expand and Factor with the trigonometric identities added by adding the option Trig->True. For example, evaluating **Expand[2Sin[x]Cos[x],Trig->True]** gives $\sin 2x$.

Section 13. Table

MATHEMATICA has a command that can iterate elements in a list. Suppose we wanted a list of the first 6 powers of 2. Evaluate **Table[2^i,{i,1,6}]** and you get Out[1]. The syntax for Table is Table[fn,{i,a,b,stp}], where *fn* is evaluated from $i = a$ to $i = b$ with step *stp*. If you omit *stp*, then the default step is 1 and a and b are assumed to be integers. Let's create a list a points and plot them. Evaluate **Table[{i,i^3},{i,0,1,.2}]** as in Out[2] of Figure 0.11. Next, evaluate **ListPlot[%]** to get the plot below In[3].

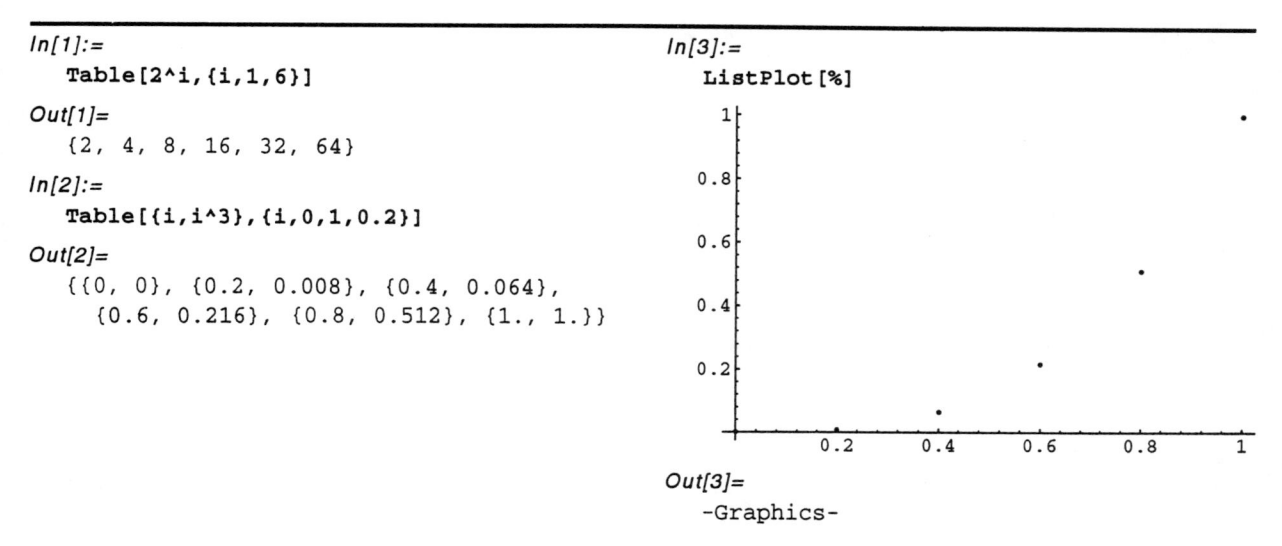

```
In[1]:=
    Table[2^i,{i,1,6}]
Out[1]=
    {2, 4, 8, 16, 32, 64}
In[2]:=
    Table[{i,i^3},{i,0,1,0.2}]
Out[2]=
    {{0, 0}, {0.2, 0.008}, {0.4, 0.064},
     {0.6, 0.216}, {0.8, 0.512}, {1., 1.}}
```

```
In[3]:=
    ListPlot[%]

Out[3]=
    -Graphics-
```

Figure 0.11: **Table command**

Section 14. Slope Fields

Suppose that we are given the derivative of function $y = f(x)$ as a function of x and $f(x)$. That is, suppose we know y' as a function of x and y. Hence, given any value of x and y, we know the slope of f and therefore we know that tangent line at that point. *MATHEMATICA* can picture these tangent lines as we vary x and y. The function PlotVectorField will do this. We must first load the package fiel.m by evaluating **<< Graphics'PlotField.m'**. If $y' = g(x, y)$, the syntax for PlotVectorField is PlotVectorField[1,g,{x,a,b},{y,c,d}], where x varies over $[a, b]$ and y varies over $[c, d]$. See the Solved Problem on Slope Fields for an example of its use.

Appendix II

Optional Files for Producing Graphics

The code in this appendix is designed to produce graphical displays of important concepts from calculus. It may appear a bit complicated and the arcane syntax might seem impossible to understand, but you do not need to understand it to use it, and we think the displays produced are worth the trouble.

Be careful to type the expressions *exactly* as they appear. Take special care to distinguish square brackets from parentheses, include all commas and semicolons, and use the brackets exactly as shown. You will only have to type this in once. After the functions are given, we will explain how to save them in a file that you can then use repeatedly. Alternatively, the file is available via ftp as explained at the end of the preface.

Functions for the MYCALC.m file

We begin with the functions `lhssum`, `rhssum`, `trap`, and `mid`, which give graphical displays of the left-hand sum, right-hand sum, trapezoidal rule, and midpoint rule. Begin with a new notebook, and clear any previous definitions. It is probably best to quit *MATHEMATICA* and restart it. Be careful to type the expressions *exactly* as they appear below.

Since we are creating a package, you may use an editor or a wordprocessor if you find this easier. We will explain how to save the file later.

```
lhssum[f_,{x_,a_,b_,n_}]:= Module[
        {delta,i,pic,plt,pts,values},
         delta = N[(b-a)/n];
         pts = Table[a+i*delta,{i,0,n}];
         values = Table[f[pts[[i]]],{i,1,n+1}];
         pic = Table[Line[{{pts[[i]],0},{pts[[i]],values[[i]]},\
             {pts[[i+1]],values[[i]]},{pts[[i+1]],0}}],{i,1,n}];
         plt = Plot[f[x],{x,a,b},DisplayFunction->Identity];
         Show[plt,Graphics[pic],DisplayFunction->$DisplayFunction]
    ]
```

```
rhssum[f_,{x_,a_,b_,n_}]:= Module[
        {delta,i,pic,plt,pts,values},
         delta = N[(b-a)/n];
         pts = Table[a+i*delta,{i,0,n}];
         values = Table[f[pts[[i]]],{i,1,n+1}];
         pic = Table[Line[{{pts[[i]],0},{pts[[i]],values[[i+1]]},\
                {pts[[i+1]],values[[i+1]]},{pts[[i+1]],0}}],{i,1,n}];
         plt = Plot[f[x],{x,a,b},DisplayFunction->Identity];
         Show[plt,Graphics[pic],DisplayFunction->$DisplayFunction]
        ]

trap[f_,{x_,a_,b_,n_}]:= Module[
        {delta,i,pic,plt,pts,values},
         delta = N[(b-a)/n];
         pts = Table[a+i*delta,{i,0,n}];
         values = Table[f[pts[[i]]],{i,1,n+1}];
         pic = Table[Line[{{pts[[i]],0},{pts[[i]],values[[i]]},\
                {pts[[i+1]],values[[i+1]]},{pts[[i+1]],0}}],{i,1,n}];
         plt = Plot[f[x],{x,a,b},DisplayFunction->Identity];
         Show[plt,Graphics[pic],DisplayFunction->$DisplayFunction]
        ]

mid[f_,{x_,a_,b_,n_}]:= Module[
        {delta,i,pic,plt,pts,values},
         delta = N[(b-a)/n];
         pts = Table[a+i*delta,{i,0,n}];
         values = Table[(f[pts[[i]]]+f[pts[[i+1]]])/2,{i,1,n}];
         pic = Table[Line[{{pts[[i]],0},{pts[[i]],values[[i]]},\
                {pts[[i+1]],values[[i]]},{pts[[i+1]],0}}],{i,1,n}];
         plt = Plot[f[x],{x,a,b},DisplayFunction->Identity];
         Show[plt,Graphics[pic],DisplayFunction->$DisplayFunction]
        ]
```

Next you should enter the following code for the function simp, which is a graphical presentation of Simpson's rule. Again, copy the function exactly.

```
simp[f_,{x_,a_,b_,n_}]:= Module[
          {delta,i,pic,plt,pts,values},
          delta = N[(b-a)/n];
          pts = Table[a+i*delta,{i,0,n}];
          values = Table[(2/3)*f[(pts[[i]]+pts[[i+1]])/2]\
                 +(f[pts[[i]]]+f[pts[[i+1]]])/6,{i,1,n}];
          pic = Table[Line[{{pts[[i]],0},{pts[[i]],values[[i]]},\
                 {pts[[i+1]],values[[i]]},{pts[[i+1]],0}}],{i,1,n}];
          plt = Plot[f[x],{x,a,b},DisplayFunction->Identity];
          Show[plt,Graphics[pic],DisplayFunction->$DisplayFunction]
              && Sum[((2/3)*f[(pts[[i]]+pts[[i+1]])/2]\
                 +(f[pts[[i]]]+f[pts[[i+1]]])/6)*delta,{i,1,n}]
       ]
```

Now enter the following code for the function **newt**, which is a graphical presentation of Newton's method:

```
newt[f_,{x_,a_,b_,start_,n_}]:= Module[
          {pt,i,pic},
          pt[0]=start;
          Do[pt[i]=pt[i-1]-f[pt[i-1]]/f'[pt[i-1]],{i,1,n}];
          pic=Table[Plot[f[pt[i]]+f'[pt[i]]*(u-pt[i]),{u,a,b},\
                 DisplayFunction->Identity],{i,0,n}];
          plt=Plot[f[x],{x,a,b},DisplayFunction->Identity];
          Show[plt,pic,DisplayFunction->$DisplayFunction]&&\
              Table[N[pt[i]],{i,0,n}]
       ]
```

Next, enter the following code for the function **euler**, which is a graphical presentation of Euler's method of approximating functions with some numerical output:

```
euler[f_,{x_,a_,b_,stp_},{y_,start_}]:= Module[
         {px,py,pic},
           px=a;
           py=start;
           pic={{a,start}};
           While[px<b,
                py=N[py+stp*f[px,py]];
                px=px+stp;
                AppendTo[pic,{px,py}]
              ];
           ListPlot[pic,PlotJoined->True]&&pic  ]
```

Finally, enter the following code for the function `eulerpic`, which is a graphical presentation of Euler's method of approximating functions without numerical output:

```
eulerpic[f_,{x_,a_,b_,stp_},{y_,start_}]:= Module[
        {px,py,pic},
        px=a;
        py=start;
        pic={{a,start}};
        While[px<b,
             py=N[py+stp*f[px,py]];
             px=px+stp;
             AppendTo[pic,{px,py}]
             ];
        ListPlot[pic,PlotJoined->True]  ]
```

Saving **MYCALC.m**: We now show you how to save this as a package, which can be used during any *MATHEMATICA* session. If you have typed the above functions in a *MATHEMATICA* notebook, you need to save it as a package. Move the mouse pointer to the icon **File**, hold the mouse button down and drag the mouse until Save as Special is highlighted. Release the button and you see a menu. Move the mouse pointer over the Package(TEXT) line and click the button. Now save the file in your work directory (or in *MATHEMATICA's* package folder) as MYCALC.m. This creates a package which you can call upon at any time.

If you are using an editor or word processor, save the file as an ASCII file called MYCALC.m. Then copy the file into you work directory for *MATHEMATICA* or into *MATHEMATICA's* package folder.

Activating **MYCALC.m**: When you are in a *MATHEMATICA* notebook and you want to use one of the functions in **MYCALC.m**, just evaluate << **MYCALC.m**. At that point, assuming you receive no error messages, you can freely use the functions defined in **MYCALC.m**.

Instructions for using **MYCALC.m** *file*: Before using any of the functions in **MYCALC.m**, you must have a function defined. For example, if $g(x) = \sin x$ has been defined by evaluating **g[x_]:= Sin[x]**, then you can use any of the functions applied to g. For example, lhssum[g,{x,a,b,n}] would make a picture of the left-hand sums for g with n partitions from $x = a$ to $x = b$. Similarly for RHSSUM, MID, TRAP, and SIMP. See the examples that follow.

Example: We will produce a picture of the midpoint sum for $f(x) = x^2$ on $[0,1]$ with $n = 10$ subintervals. Evaluate **Clear[f]** and **f[x_]:= x^2**. Now load MYCALC.m by evaluating << **MYCALC.m**. Evaluating **mid[f,{x,0,1,10}]** yields Out[4] of Figure 0.12. The syntax of mid is mid[f,{x,a,b,n}], which pictures f on $[a, b]$ with n subintervals. The syntax of lhssum and rhssum are the same.

```
In[1]:=
   Clear[f]
In[2]:=
   f[x_]:=x^2
In[3]:=
   <<MYCALC.m
In[4]:=
   mid[f,{x,0,1,10}]
```

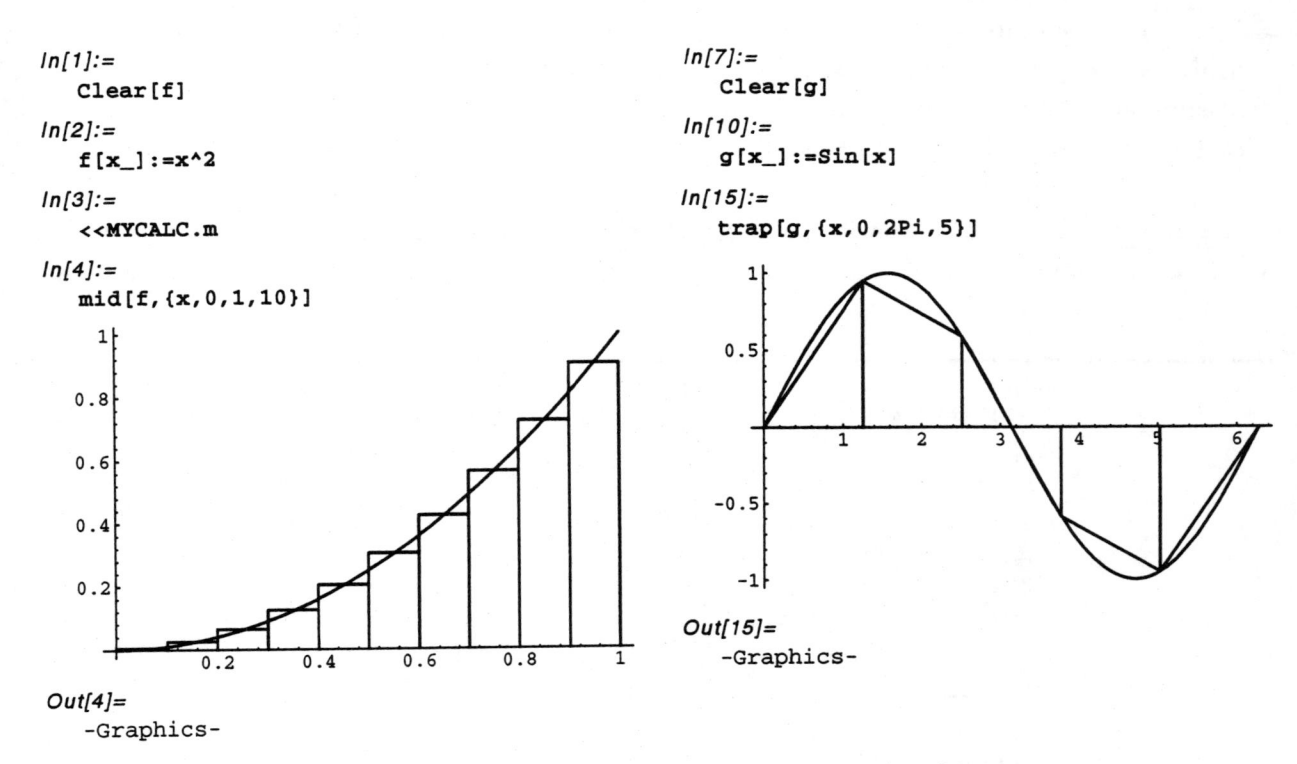

Out[4]=
   -Graphics-

```
In[7]:=
   Clear[g]
In[10]:=
   g[x_]:=Sin[x]
In[15]:=
   trap[g,{x,0,2Pi,5}]
```

Out[15]=
   -Graphics-

Figure 0.12: **A picture of midpoint and trapezoid sums**

Example: We will produce a picture of the trapezoid sum for $g(x) = \sin x$ on $[0, 2Pi]$ with $n = 5$. Evaluate **Clear[g]** and **g[x_]:= Sin[x]**. Assuming we have loaded MYCALC.m, we evaluate **trap[g,{x,0,Pi,5}]** to get Out[15] of Figure 0.12. The syntax for trap is trap[f,{x,a,b,n}], which pictures the trapezoidal sum for f on $[a, b]$ with n subintervals.

Example: We will make a picture of Simpson's rule for $f(x) = \sin x$ on the interval $[-2\pi, 2\pi]$ using $n = 6$ subintervals. First, evaluate **Clear[f]** and **f[x_]:= Sin[x]**. Then evaluate **simp[f,{x,-2Pi,2Pi,6}]**, which will produce the picture above Out[4] of Figure 0.13. Notice that simp also shows the sum that is calculated. The syntax of simp is simp[f,{x,a,b,n}], which pictures Simpson's rule for f on $[a, b]$ with n subintervals.

Example: In Figure 0.13 we will show 4 iterations of Newton's method applied to $h(x) = x^2 - 2$ with starting point $x_0 = 3$. Evaluate **Clear[h]** and **h[x_]:= x^2-2**. We choose an interval around

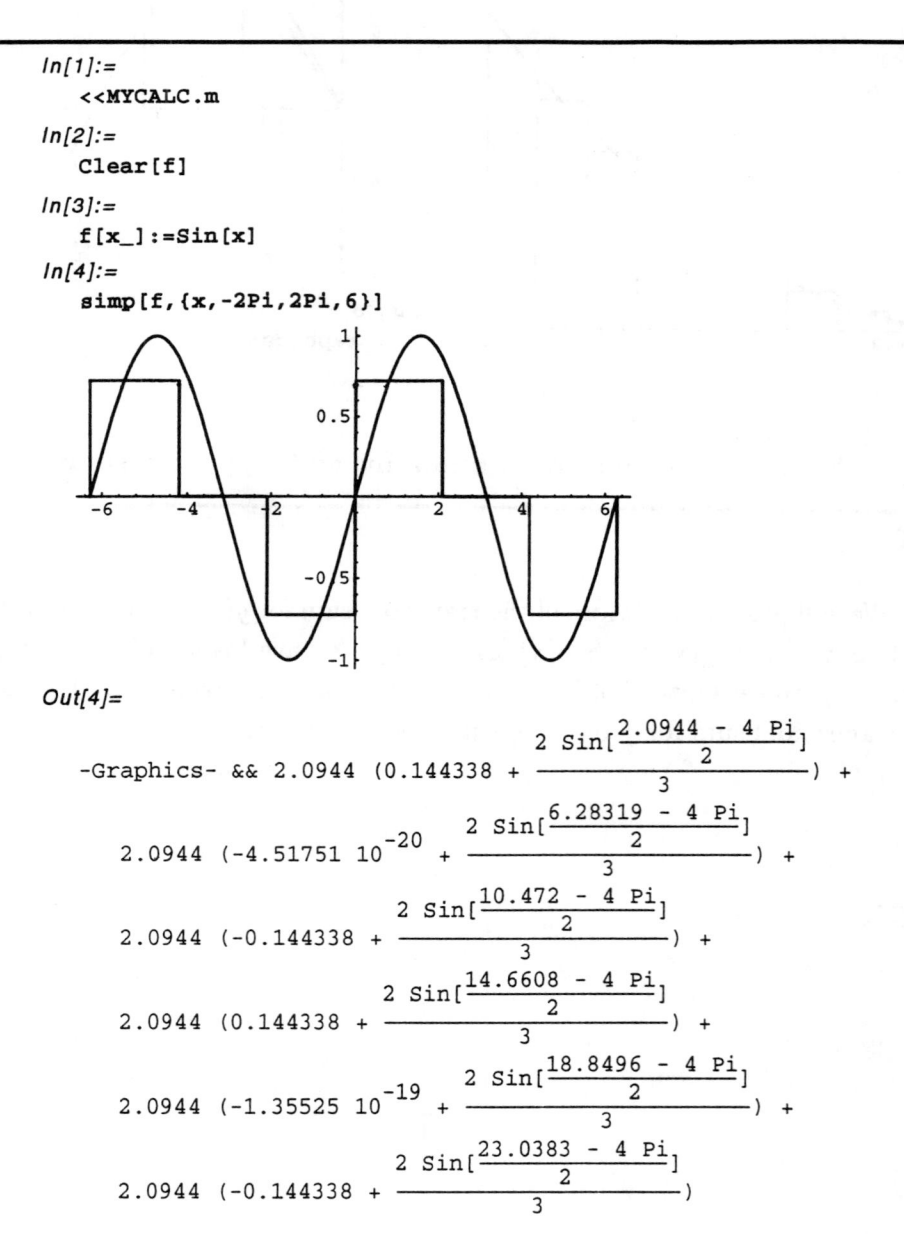

```
In[1]:=
    <<MYCALC.m
In[2]:=
    Clear[f]
In[3]:=
    f[x_]:=Sin[x]
In[4]:=
    simp[f,{x,-2Pi,2Pi,6}]
```

Out[4]=

$$-\text{Graphics}- \;\&\&\; 2.0944 \left(0.144338 + \frac{2\sin[\frac{2.0944 - 4\text{ Pi}}{2}]}{3}\right) +$$

$$2.0944 \left(-4.51751 \cdot 10^{-20} + \frac{2\sin[\frac{6.28319 - 4\text{ Pi}}{2}]}{3}\right) +$$

$$2.0944 \left(-0.144338 + \frac{2\sin[\frac{10.472 - 4\text{ Pi}}{2}]}{3}\right) +$$

$$2.0944 \left(0.144338 + \frac{2\sin[\frac{14.6608 - 4\text{ Pi}}{2}]}{3}\right) +$$

$$2.0944 \left(-1.35525 \cdot 10^{-19} + \frac{2\sin[\frac{18.8496 - 4\text{ Pi}}{2}]}{3}\right) +$$

$$2.0944 \left(-0.144338 + \frac{2\sin[\frac{23.0383 - 4\text{ Pi}}{2}]}{3}\right)$$

In[5]:=
```
Clear[h]
```
In[6]:=
```
h[x_]:=x^2-2
```
In[11]:=
```
newt[h,{x,1.2,3.1,3,4}]
```

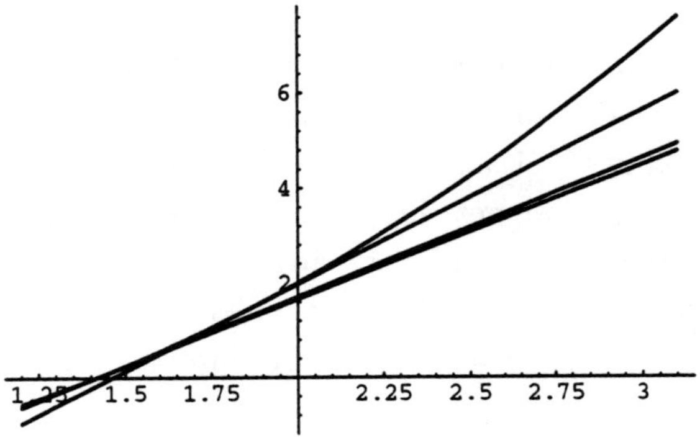

Out[11]=
   -Graphics- && {1.83333, 1.46212, 1.415, 1.41421}

In[12]:=
```
Clear[f]
```
In[13]:=
```
f[x_,y_]:=x^2+y^2
```
In[17]:=
```
euler[f,{x,0,.6,0.2},{y,1}]
```

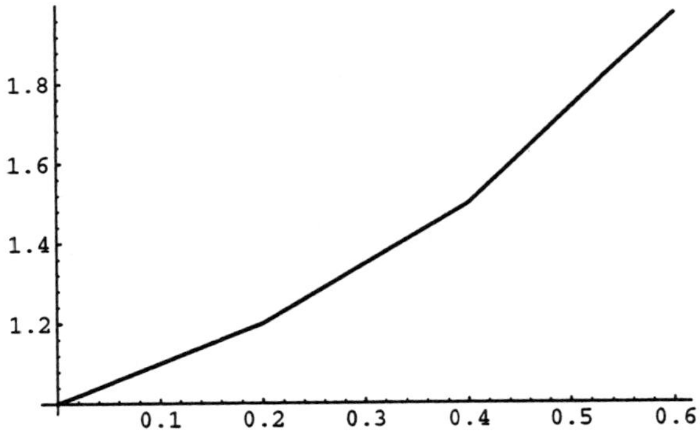

Out[17]=
   -Graphics- && {{0, 1}, {0.2, 1.2}, {0.4, 1.496},
      {0.6, 1.9756}}

Figure 0.13: **A picture of Simpson's rule, Newton's method, and Euler's method**

313

$x_0 = 3$, say [1.5, 3.2]. Next, evaluate **newt[h,{x,1.5,3.2,3,4}]**. This will produce the picture above Out[11] of Figure 0.13. Notice that below the graph, the values that Newton's method yield are displayed. They are {1.83333, 1.46211, 1.415, 1.41421}. The last one is very close to $\sqrt{2}$. The syntax of newt is newt[f,{x,a,b,start,n}], which pictures Newton's method on the interval $[a, b]$ for f, starting at $x = start$ for n iterations.

Example: In Figure 0.13 we picture Euler's method for approximating the function $y = f(x)$, if $f(0) = 1$ given $f'(x) = x^2 + y^2$. We will find this approximation on $[0, 0.6]$ using steps of size 0.2. Euler's method is described in CCH 9.3. Evaluate **Clear[f]** and **f[x_,y_]:= x^2+y^2**. Evaluating **euler[f,{x,0,0.6,0.2},{y,1}]**, we get the picture above Out[17] i Figure 0.13. Note that euler also returns the points that are used in the approximation of the function. If you only want the graph, use EULERPIC. The syntax of euler is euler[f,{x,a,b,step},{y,start}], which pictures an approximation of the function $y = g(x)$ on the interval $[a, b]$ if $g' = f(x, y)$ and $g(a) = start$.

The syntax for eulerpic is the same.

Index of Solved Problems

Chapter 1

Domains, ranges, and zeros of functions (CCH Text 1.1) 1

Testing exponential data (CCH Text 1.3) 9

Powers versus exponentials (CCH Text 1.4) 16

Inverses of functions (CCH Text 1.5) 21

Properties of logarithms from graphs (CCH Text 1.6) 31

Approximating the number e (CCH Text 1.7) 35

Periodicity of trigonometric functions (CCH Text 1.10) 49

Rational functions (CCH Text 1.11) 55

Vertical asymptotes (CCH Text 1.11) 56

Chapter 2

Calculating velocities (CCH Text 2.1) 61

Average rates of change (CCH Text 2.2) 69

Instantaneous rates of change (CCH Text 2.2) 70

The derivative function (CCH Text 2.3) 83

Calculating limits graphically (CCH Text 2.7) 89

Chapter 3

Measuring distance (CCH Text 3.1) 93

Calculating Riemann sums (CCH Text 3.2) 99

Limits of Riemann sums (CCH Text 3.2) 102

Chapter 4

Difference quotients and derivatives (CCH Text 4.2) 119

The product rule (CCH Text 4.4) 123

Implicit differentiation (CCH Text 4.8) 127

Chapter 5

Maxima and minima (CCH Text 5.1) 133

Critical points and extrema (CCH Text 5.1) 135
Inflection points (CCH Text 5.2) 139
Welding boxes (CCH Text 5.6) 151
Newton's method (CCH Text 5.7) 159

Chapter 6

Families of antiderivatives (CCH Text 6.3) 163

Chapter 7

Integrating with MATHEMATICA (CCH Text 7.1) 169
Implementing the right-hand rule (CCH Text 7.6) 173
Implementing the midpoint rule (CCH Text 7.6) 174
Calculating improper integrals (CCH Text 7.8) 185
Approximating improper integrals (CCH Text 7.9) 193
Approximating improper integrals II (CCH Text 7.9) 199

Chapter 8

An oil slick (CCH Text 8.1) 205
Area and center of mass (CCH Text 8.2) 207
Arc length, area, and volume (CCH Text 8.2) 213
From the Earth to the moon (CCH Text 8.3) 221
Rainfall in Anchorage (CCH Text 8.6) 225

Chapter 9

Families of solutions (CCH Text 9.1) 229
Slope fields (CCH Text 9.2) 233
Euler's method (CCH Text 9.3) 237
A tank of water (CCH Text 9.6) 241
Terminal velocity (CCH Text 9.6) 247
Population growth with a threshold (CCH Text 9.7) 251
Spider mites and ladybugs (CCH Text 9.8) 257
Springs and second-order equations (CCH Text 9.10) 261

Chapter 10

Approximating $\cos x$ **(CCH Text 10.1)** 265

Interval of convergence (CCH Text 10.2) 273

The error in Taylor approximations (CCH Text 10.4) 279

Fourier series (CCH Text 10.5) 283

Index of Laboratory Exercises

Chapter 1

Zeros, Domains, and Ranges (CCH Text 1.1)..............................7
Fitting Exponential Data (CCH Text 1.3)................................11
U. S. Census Data (CCH Text 1.3).......................................13
Powers versus Exponentials (CCH Text 1.4)..............................19
The Inverse of a Function (CCH Text 1.5)...............................25
Restricting the Domain (CCH Text 1.5).................................27
The Inverse of an Exponential Function (CCH Text 1.5).................29
Seeing Log Identities Graphically (CCH Text 1.6)......................33
Approximating e (CCH Text 1.8).......................................39
Seeing Log Identities Graphically II (CCH Text 1.7)...................41
Growth Rates of Functions (CCH Text 1.7)..............................43
A Graphical Look at Borrowing Money (CCH Text 1.8)...................45
Shifting and Stretching (CCH Text 1.9)................................47
Equations Involving Trig Functions (CCH Text 1.10)...................51
Inverse Trig Functions (CCH Text 1.10)................................53
Asymptotes (CCH Text 1.11)..59

Chapter 2

Average Velocity (CCH Text 2.1).......................................65
Falling with a Parachute (CCH Text 2.1)...............................67
Slopes and Average Rates of Change (CCH Text 2.1).....................75
Tangent Lines and Rates of Change (CCH Text 2.2).....................77
The Derivative of the Gamma Function (CCH Text 2.2)..................79
The Meaning of the Sign of f' (CCH Text 2.3)........................85
Recovering f from f' (CCH Text 2.5)...............................87
Using Graphs to Estimate Limits (CCH Text 2.7).......................91

Chapter 3

An Asteroid (CCH Text 3.1) 95
A Falling Water Table (CCH Text 3.1) 97
Calculating Riemann Sums (CCH Text 3.2) 105
Limits of Riemann Sums (CCH Text 3.2) 107
Estimating Integrals with Riemann Sums (CCH Text 3.2) 109
Riemann Sums and the Fundamental Theorem (CCH Text 3.4) 111
Calculating Areas (CCH Text 3.3) 115
The Average Value of a Function and the Fundamental Theorem
  (CCH Text 3.4) 117

Chapter 4

Difference Quotients and the Derivative (CCH Text 4.2) 121
The Quotient Rule and Chain Rule (CCH Text 4.5) 125
Implicit Differentiation (CCH Text 4.8) 131

Chapter 5

Local Extrema (CCH Text 5.1) 137
Inflection Points (CCH Text 5.2) 141
The George Deer Reserve (CCH Text 5.2) 143
The Meaning of the Signs of f' and f'' (CCH Text 5.2) 145
Families of Curves (CCH Text 5.3) 149
Building Boxes (CCH Text 5.6) 153
Building Fuel Tanks (CCH Text 5.6) 155
Roads (CCH Text 5.6) 157
Newton's Method (CCH Text 5.7) 161

Chapter 6

Antiderivatives of $\arctan x$ (CCH Text 6.3) 165
An Antiderivative of $\sin(x^2)$ (CCH Text 6.3) 167

Chapter 7

    Integration with *MATHEMATICA* (CCH Text 7.1).........................171
    Implementing the Left-Hand Rule (CCH Text 7.6).......................177
    Implementing the Trapezoidal Rule and Simpson's Rule (CCH Text 7.7)181
    The Trapezoidal Rule with Error Control (CCH Text 7.7)................183
    Evaluating Improper Integrals (CCH Text 7.8).............................189
    The Gamma Function (CCH Text 7.8) 191
    Approximating Improper Integrals (CCH Text 7.9).......................197
    Approximating Improper Integrals II (CCH Text 7.9) 203

Chapter 8

    An Oil Spill in the Ocean (CCH Text 8.1).................................209
    A Center of Mass of a Sculpture (CCH Text 8.1).........................211
    Arc Length and Volume (CCH Text 8.2)...................................215
    Arc Length and Limits (CCH Text 8.2) 217
    Relation of Arc Length to Area (CCH Text 8.2).........................219
    From the Earth to the Sun (CCH Text 8.3)..............................223
    SAT Scores (CCH Text 8.6) 227

Chapter 9

    Families of Solutions (CCH Text 9.1).......................................231
    Slope Fields (CCH Text 9.2)..235
    Estimating Function Values with Euler's Method (CCH Text 9.3).......239
    A Leaky Balloon (CCH Text 9.5)...243
    Baking Potatoes (CCH Text 9.5)...245
    Drag and Terminal Velocity (CCH Text 9.6)..............................249
    Comparing Population Models (CCH Text 9.7)...........................255
    Foxes and Hares (CCH Text 9.8)...259
    A Damped Spring (CCH Text 9.11) 263

Chapter 10

> Taylor Polynomials and the Cosine Function: The Expansion Point (CCH Text 10.1)...269
>
> Taylor Polynomials and the Cosine Function: The Degree (CCH Text 10.2)...271
>
> Interval of Convergence (CCH Text 10.2).................................275
>
> Approximating π (CCH Text 10.3)...277
>
> Approximating $\sec\frac{1}{2}$ (CCH Text 10.5)...281
>
> Fourier Approximations for $|x|$ (CCH Text 10.6).........................285